Impacts of Climate and Human Activities on Water Resources and Quality

Weili Duan · Kaoru Takara

Impacts of Climate and Human Activities on Water Resources and Quality

Integrated Regional Assessment

Springer

Weili Duan
State Key Laboratory of Desert
and Oasis Ecology
Xinjiang Institute of Ecology
and Geography, Chinese Academy
of Sciences
Urumqi, Xinjiang, China

Kaoru Takara
The Graduate School of Advanced
Integrated Studies in Human Survivability
Kyoto University
Kyoto, Japan

This book is jointly supported by the National Natural Science Foundation of China (Grant No. 41501552), the National Thousand Talents Program (Grant No. Y971171), and the Pioneer "Hundred Talents Program" of the Chinese Academy of Sciences (Grant No. 2018-126), as well as the "Japan-ASEAN Science, Technology and Innovation Platform (JASTIP): Promotion of Sustainable Development Research" within the framework of the Collaboration Hubs for International Research Program (CHIRP) 2015 funded by the Strategic International Collaborative Research Program (SICORP) of Japan Science and Technology Agency (JST).

ISBN 978-981-13-9393-8　　ISBN 978-981-13-9394-5 (eBook)
https://doi.org/10.1007/978-981-13-9394-5

This Springer imprint is published by the registered company Springer Nature Singapore Pte Ltd.
The registered company address is: 152 Beach Road, #21-01/04 Gateway East, Singapore 189721, Singapore

Foreword

With the rapid increase of world population, the global water shortage is set to be the major crises of the twenty-first century, that is, population dynamics (growth, age distribution, urbanization, and migration) create pressures on freshwater resources due to the increased water demands and pollution. Moreover, water resource management faces a new challenge, i.e., the potential for longer term and more persistent climate change nowadays, which, in coming years, may significantly affect the availability of supply and patterns of water demand. Evidence for climate change impacts on the hydro-climatology of Asia is plentiful. The possibility of frequent occurrence of extremely low rainfall, decrease in snowfall, and earlier thaw will tend to increase the vulnerability of water resources. Meanwhile, extreme rainfall and temperature tend to cause hydrological disasters including floods, water quality incidents, and so on.

This new book *Impacts of Climate and Human Activities on Water Resources and Quality—Integrated Regional Assessment* is of course the biggest success, in which the authors put great efforts and invested long time on water-related social–environmental changes in China, Japan, and the Aral Sea Basin. All these regions are located at the similar ranges of latitude, but have different climate types, economic development stages, and water-related problems. Based on multiple datasets and methods, the authors have successfully expounded water related social–environmental changes in these regions.

All in all, the results obtained in this book substantially enhance the knowledge of climate change impacts on the water quality and water resources in Asian countries, which may provide a means to reduce water quality incidents and mitigate future impacts by adapting water management. Furthermore, the improved methods for water quality modeling in data-scarce regions are potentially transferable to other study areas and applicable in future research. I trust that this book will provide a useful knowledge base and tool for researchers and resource managers to improve water resource management.

January 2020

Prof. Chong-Yu Xu
Fellow of Norwegian Academy of Science and Letters
Fellow of Norwegian Academy of Technological Sciences
University of Oslo
Oslo, Norway

Acknowledgements

We would like to express our sincere gratitude and appreciation to collaborators from Chinese Academy of Sciences and Kyoto University for their attention and continuous support. Also, this book is jointly supported by the National Thousand Talents Program (Grant No. Y971171), the Pioneer "Hundred Talents Program" of the Chinese Academy of Sciences (Grant No. 2018-126), and the Program for Foreign High-Level Talents Introduction in Xinjiang Uygur Autonomous Region (Y941181), as well as the "Japan-ASEAN Science, Technology and Innovation Platform (JASTIP): Promotion of Sustainable Development Research" within the framework of the Collaboration Hubs for International Research Program (CHIRP) 2015 funded by the Strategic International Collaborative Research Program (SICORP) of Japan Science and Technology Agency (JST).

Weili Duan
Kaoru Takara

Parts of this monograph have been published in the following journal articles:

Weili Duan, Bin He, Daniel Nover, Jingli Fan, Guishan Yang, Wen Chen, Huifang Meng, Floods and associated socioeconomic damages in China over the last century, Natural Hazards, 82 (1): 401–413, 2016.

Weili Duan, Bin He, Daniel Nover, Guishan Yang, Wen Chen, Huifang Meng, Shan Zou, Chuanming Liu. Water Quality Assessment and Pollution Source Identification of the Eastern Poyang Lake Basin Using Multivariate Statistical Methods, Sustainability, 8(2): 133, 2016.

Weili Duan, Bin He, Kaoru Takara, Pingping Luo, Daniel Nover, Netrananda Sahu, Yosuke Yamashiki, Spatiotemporal evaluation of water quality incidents in Japan between 1996 and 2007. Chemosphere, 93(6): 946–953. 2013.

Weili Duan, Bin He, Kaoru Takara, Pingping Luo, Maochuan Hu, Nor Eliza Alias, Daniel Nover. Changes of precipitation amounts and extremes over Japan between 1901 and 2012 and their connection to climate indices, Climate Dynamics, 45(7): 2273–2292, 2015.

Duan, W., He, B., Chen, Y., Zou, S., Wang, Y., Nover, D., Chen, W., Yang, G. Identification of long-term trends and seasonality in high-frequency water quality data from the Yangtze River basin, China. Plos One, 13, e188889, 2018.

Weili Duan, Bin He, Netrananda Sahu, Pingping Luo, Daniel Nover, Maochuan Hu, and Kaoru Takara. Spatiotemporal variability of Hokkaido's seasonal precipitation in recent decades and connection to water vapor flux, International Journal of Climatology, 37 (9), 3660–3673, 2017.

Weili Duan, Kaoru Takara, Bin He, Pingping Luo, Daniel Nover, Yosuke Yamashiki, Spatial and temporal trends in estimates of nutrient and suspended sediment loads in the Ishikari River, Japan, 1985 to 2010, Science of the Total Environment, 461: 499–508, 2013.

Weili Duan, Kaoru Takara, Bin He, Pingping Luo, Daniel Nover, and Maochuan Hu. Modeling suspended sediment sources and transport in the Ishikari River basin, Japan using SPARROW, Hydrology and Earth System Sciences, 19, 1293–1306, 2015.

Weili Duan Bin He, Kaoru Takara, Pingping Luo, Daniel Nover, and Maochuan Hu. Impacts of climate change on the hydro-climatology of the upper Ishikari river basin, Japan. Environmental Earth Sciences, 76 (14): 490, 2017.

Weili Duan, Yaning Chen, Shan Zou, and Daniel Nover. Managing the water-climate-food nexus for sustainable development in Turkmenistan. Journal of Cleaner Production, 220: 212–224, 2019.

Contents

Chapter 1
Introduction

1.1 Background

The sustainable development of human social society and economy requires access to water resources. With the rapid growth of the world's population, global water shortages will become the main crisis of the twenty-first century (Eliasson 2015; Pearce 2006; Schade and Pimentel 2010; Vörösmarty et al. 2010); more specifically, population dynamics (growth, age distribution, urbanization, and migration) create pressures on freshwater resources due to the increased water demands and pollution (Mekonnen and Hoekstra 2016; Polizzotto et al. 2008). At present, climate change is also a big problem to increase the burden of water resources (Arnell 1999; Barnett et al. 2005) and energy security (Behrens et al. 2017). According to the World Health Organization (WHO)'s report, about 1.2 billion people (almost one-fifth of the world's population) live in areas of physical scarcity, and about 500 million people are in this situation; another about 1.6 billion people (almost one-quarter of the world's population) face economic water shortage (where countries lack the necessary infrastructure to take water from rivers and aquifers) (Fig. 1.1).

Due to effects of climate change, water-related disasters arising from floods, droughts, tropical cyclones, landslides, and tsunamis are undoubtedly increased over recent decades and appear to continue to rising (Wu et al. 2019), which pose major impediments to achieving human security and sustainable socio-economic development (Horne et al. 2018; Milly et al. 2002), as recently witnessed with disasters such as Hurricane Katrina in 2005 (Brunkard et al. 2008), 2011 Tohoku earthquake and tsunami (Fujii et al. 2011), and Thai Floods 2011 (Haraguchi and Lall 2015). Figure 1.2 shows changes of the world water-related disaster events, suggesting an increasing trend from 1970 to 2013. From 1970 to 2013, floods and storms increased dramatically, but during this period, other types of disasters did not increase significantly. The average flood from 2001 to 2013 was twice the average of 1986–2000, and the storm increased more than 1.5 times.

The interaction of extreme weather and climate events with exposed and fragile humans and natural systems can lead to disasters that often cause significant damage

W. Duan and K. Takara, *Impacts of Climate and Human Activities on Water Resources and Quality*, https://doi.org/10.1007/978-981-13-9394-5_1

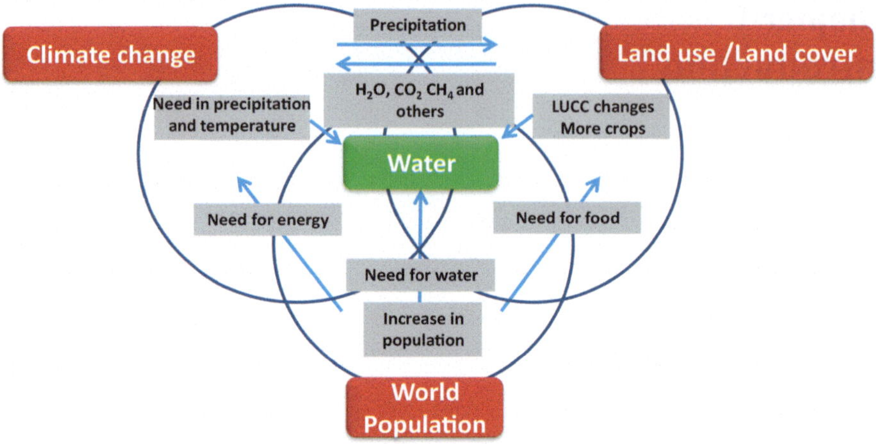

Fig. 1.1 Drivers of global change and feedbacks to the global water cycle. (Modified from the Harding and Kabat 2007)

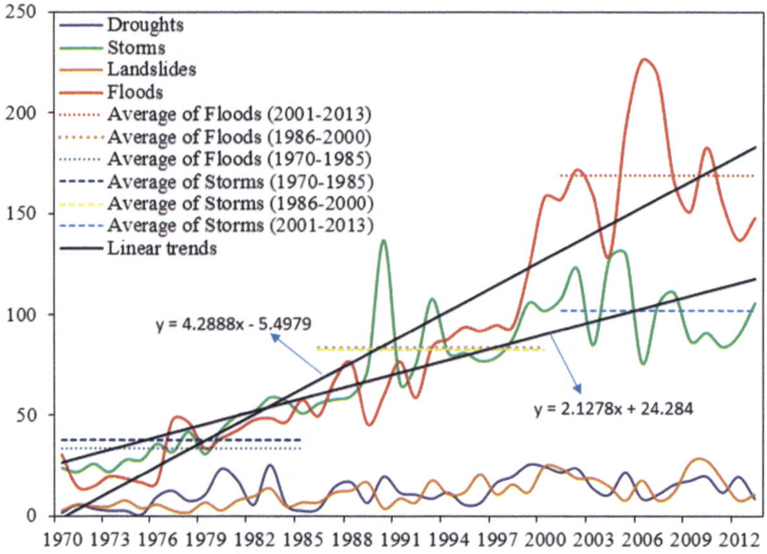

Fig. 1.2 Changes of water-related disaster events over the world from 1970 to 2013. (Data from the Emergency Event Database of the Centre for Research on the Epidemiology of Disasters. http://www.emdat.be)

to property and casualties. During the period of 1970–2013, 11,707 hydrological disasters reported worldwide killed 3,525,166 people, resulting in a record loss of US$ 2,600,128,516 of damages (Table 1.1).

Table 1.1 Hydrological disasters from 1970 to 2013

Continent	Occurrence	Deaths	Injured	Affected	Damage (000 US$)
Africa	2,262	859,176	81,758	452,256,918	26,574,852
Americas	2,819	519,718	2,938,072	236,704,736	948,201,308
Asia	4,567	1,953,106	3,559,008	5,860,665,083	1,195,187,529
Europe	1,523	186,681	85,362	38,520,962	356,116,657
Oceania	536	6,485	9,928	20,609,528	74,048,170
Total	11,707	3,525,166	6,674,128	6,608,757,227	2,600,128,516

Source Frequencies are authors' estimates based on data from the Emergency Event Database of the Centre for Research on the Epidemiology of Disasters. http://www.emdat.be

Figure 1.3 more clearly shows damages of the human and economic impact by disaster types in 2013 compared to the average 2003–2012. Clearly, the damage caused by extreme weather and climate events accounted for a very large part of total disasters. In addition, we can see most of the disasters occurred in Asia. For example, the number of disasters happened in 2013 in Asia was up to approximately 88.16%.

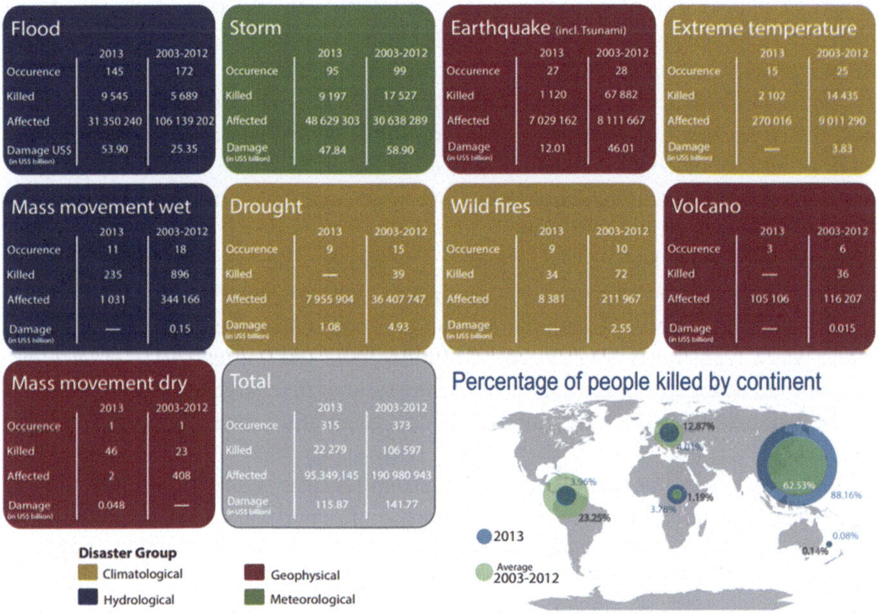

Fig. 1.3 Human and economic impact by disaster types (2013 versus average 2003–2012) (*Source* The Emergency Event Database of the Centre for Research on the Epidemiology of Disasters. http://www.emdat.be)

Meanwhile, water pollution caused by human activities and natural activities is another serious problem that increases pressures on freshwater resources supply. An increase in population will exacerbate land use (DeFries et al. 2010; Zou et al. 2019), not only causing changes in evaporation and runoff, but also causing changes in greenhouse gas emissions (Rosa and Dietz 2012). Water quality is influenced by many factors such as nutrients, sedimentation, temperature, pH, heavy metals, non-metallic toxins, persistent organics and pesticides, and biological factors (Carr and Neary 2008; Duan et al. 2018). Firstly, increasing population will require an increase in agricultural productivity, which will naturally result in increased irrigation backflow due to the use of fertilizers and pesticides. Secondly, changes in land use caused by population dynamics will greatly affect the aquatic environment. For example, deforestation will increase since more cropland and wood for fuel are needed, accelerating erosion and leaching and increasing water pollution. Thirdly, climate change has a major impact on the world's freshwater resources, water quality, and water management (Bates et al. 2008; Pachauri 2008). The increase in surface temperature caused by global warming and the changes in runoff time and runoff caused by changes in space-time precipitation are likely to produce unfavorable changes in surface water quality, which in turn affects the health of human ecosystems. For example, higher surface water temperatures may increase biological productivity, increase the amount of bacteria and fungi in the water, and promote algal blooms since temperature can control the types of aquatic life that can survive, regulates the amount of dissolved oxygen in the water, and influences the rate of chemical and biological reactions (Kundzewicz et al. 2008). Increased flooding from extreme rainfall and periodic storm surges caused by rising sea levels may affect water quality, overloading infrastructure, such as stormwater drainage operations, wastewater systems, treatment facilities, mine tailing impoundments, and landfills, which probably increase the risk of contamination (Duan et al. 2013; McCarthy 2001). In general, deterioration of water quality caused by climate change and human activity needs to be stopped for a sustainable society.

Figure 1.4 shows that the driver scores for human water security (HWS) threats and biodiversity (BD) threats to river systems, which indicates that nutrients such as nitrogen and phosphorus, sediment, and human population are the main factors in human water security (HWS) threats and biodiversity (BD) threats to river systems. Here, human water stress was calculated as the ratio of discharge to local population, thereby capturing the negative impact of high population density and low natural water supply.

Moreover, water managers face new uncertainties, that is, the potential for longer term and more persistent climate change may significantly affect the availability of supply and patterns of water demand (Vörösmarty et al. 2000). Climate change is a major and lasting change in the statistical distribution of weather patterns over periods ranging from decades to millions of years, which affects the hydrological cycle and thus affects users' water use. Firstly, the distribution of precipitation in space and time is very uneven, leading to huge temporal variability in water resources worldwide (Oki and Kanae 2006). For example, the Atacama Desert in Chile, the driest place on earth, receives imperceptible annual quantities of rainfall each year. On

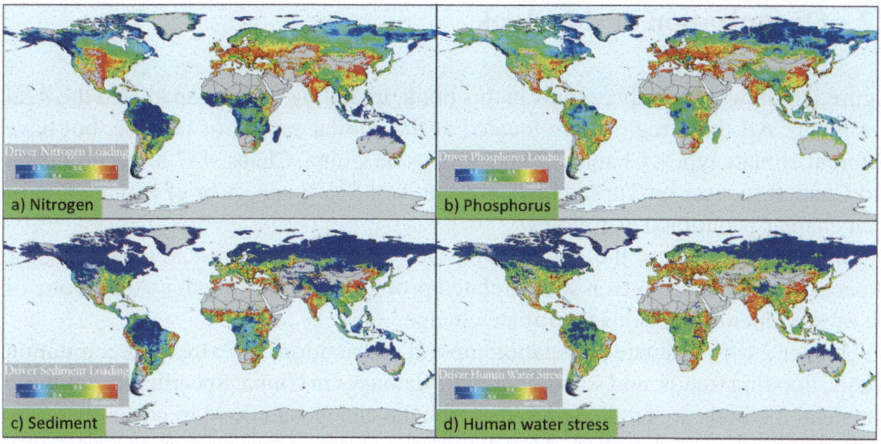

Fig. 1.4 Global geography of driver scores for human water security (HWS) threats and biodiversity (BD) threats to river system: **a** nitrogen loading, **b** phosphorus loading, **c** sediment loading, and **d** human water stress. Maps display standardized scores, indicating the spatial distribution of relative threats from low (blue) to high (red). (Data from Vörösmarty et al. 2010)

the other hand, Mawsynram, Assam, India receives more than 450 inches annually. If all the freshwater on the planet were divided equally among the global population, there would be 5,000–6,000 m^3 of water available per person each year (Vörösmarty et al. 2000). Secondly, the rate of evaporation varies a great deal, depending on temperature and relative humidity, which affects the amount of water that can be used to recharge groundwater. Shorter duration but greater rainfall (meaning more runoff and less infiltration) plus a combination of evapotranspiration (sum of evaporation from the surface to the atmosphere and plant transpiration) and increased irrigation are expected to lead to depletion of groundwater (Konikow and Kendy 2005; Wada et al. 2010).

All in all, global change including all the future changes due to anthropogenic activities has significantly affected water resources availability and quality, extreme events, surface and groundwater, marine and continental water. Therefore, it is momentous to evaluate the water quality, extreme events, and water resources under climate change, which can help us to fully understand the relationship between the hydrological process and the watershed characters and identify the trend of the vulnerable of the water-related problems. Also, the above analysis has clearly demonstrated that Asia is one of the most vulnerable regions to climate change. Therefore, the objective of this book is to evaluate the water-related disasters, precipitation extremes, and water resource managements in Asian countries under climate change.

1.2 Organization of this Book

Figure 1.5 shows the study regions in this book, including China, Japan, and the Aral Sea Basin. All these regions are located at the similar ranges of latitude, but have different climate types. Chapters 2 and 3 focus on China, Chaps. 4–7 focus on Japan, and Chap. 8 focuses on Turkmenistan of the Aral Sea Basin. A brief explanation of each chapter is outlined as follows.

This chapter depicts the background and concludes the research progress of impacts of climate change on water-related problems, suggesting that Asia is one of the most vulnerable regions to climate change.

Chapter 2 is to evaluate the spatiotemporal distributions of extreme precipitation events, flooding trends, and socio-economic damages in China. Specifically, multiple statistical methods are employed to describe the changes in flood trends and associated socio-economic damages. Also, possible causes such as extreme precipitation events are explored and discussed.

Chapter 3 uses different multivariate statistical techniques including cluster analysis, discriminant analysis, and component analysis/factor analysis to evaluate the temporal and spatial variation of surface water quality in the lake basin of China, based on water quality data at 28 sites from January 2012 to April 2015.

Chapter 4 gives a spatiotemporal evaluation of water quality incidents in Japan to provide numerous insights into incident numbers, pollutant category, incident cause, and affected extend. Results show that human activities and extreme events

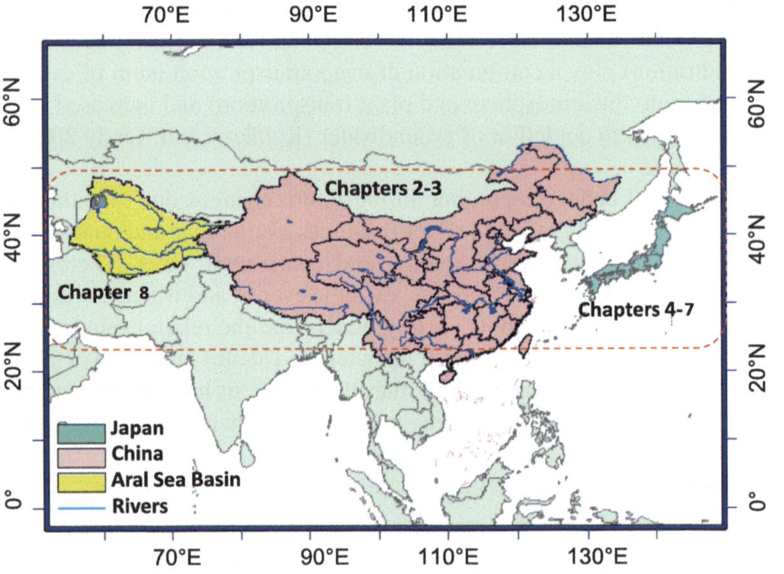

Fig. 1.5 Locations of the study regions, including China, Japan, and the Aral Sea Basin

(e.g., extreme precipitation) were the causes and have significant implications for adaptation measures, strategies, and policies to reduce water quality incidents.

Chapter 5 evaluates changes of the annual and seasonal precipitation amounts and annual precipitation extreme indices in Japan by calculating the Expert Team on Climate Change Detection and Indices (ETCCDI) at 51 weather stations from 1901 to 2012. Possible correlation with climate indices and sea surface temperatures (SSTs) near Japan was explored. Results show the Hokkaido area was a special area for precipitation extremes, suggesting further studies are needed in Hokkaido.

Chapter 6 characterizes annual, seasonal, and monthly precipitation trends in Hokkaido during the period 1980–2011 using the Mann–Kendall test, Moran's I, and geostatistical interpolation techniques. Possible association with large-scale atmospheric circulation was explored using NCEP/NCAR reanalysis data. The changes in precipitation can contribute to the following chapters. Using this historical datasets and the Soil and Water Assessment Tool (SWAT) model, this chapter then simulates the possible effects of climate change on water resources for the 2030s, 2060s, and 2090s periods in the Upper Ishikari River Basin based on the General Circulation Model (GCM) outputs and the Statistical DownScaling Model (SDSM). Uncertainties were also discussed about the GCM, the downscaling method and the hydrologic model.

Chapter 7 develops regression equations by combining the Maintenance of Variance-Extension type 3 (MOVE. 3) and the regression model Load Estimator (LOADEST) to estimate total nitrogen (TN), total phosphorus (TP), and suspended sediment (SS) loads in the Ishikari River, Japan, based on water quality data at five sites from January 1985 to December 2010. Based on these statistical results, this chapter also develops a SPARROW-based suspended sediment (SS) model for surface waters in the Ishikari River Basin, the largest watershed in Hokkaido, based on water quality monitoring records at 31 stations during the period 1982–2010. The model contained four source variables including developing lands, forest lands, agricultural lands, and stream channels, three landscape delivery variables including slope, soil permeability, and precipitation, two in-stream loss coefficients including small stream (drainage area ≤ 200 km^2) and big stream (drainage area > 200 km^2), and reservoir attenuation. Results obtained in this chapter can help resource managers identify priority sources of pollution and mitigate water pollution in order to safeguard water resources and protect aquatic ecosystems.

Chapter 8, firstly, analyzes climate change impacts on water resources in the Aral Sea Basin, and then employs climate predictions to analyze the water balance for the Amu Darya River Basin and evaluate future water use, crop yields, and land and water productivities in Turkmenistan from 2016 to 2055. Results can provide insights to help resource managers to fully understand the Aral Sea Basin crisis and identify vulnerabilities in water–land–climate nexus to ensure food security, water management, and sustainable development.

References

Arnell, N. W. (1999). Climate change and global water resources. *Global Environmental Change, 9,* S31–S49.

Barnett, T. P., Adam, J. C., & Lettenmaier, D. P. (2005). Potential impacts of a warming climate on water availability in snow-dominated regions. *Nature, 438,* 303.

Bates, B., Kundzewicz, Z. W., Wu, S., & Palutikof, J. (2008). Climate change and water, Intergovernmental Panel on Climate Change (IPCC). Technical Paper of the Intergovernmental Panel on Climate Change, IPCC Secretariat, Geneva (p. 210).

Behrens, P., Van Vliet, M. T., Nanninga, T., Walsh, B., & Rodrigues, J. F. (2017). Climate change and the vulnerability of electricity generation to water stress in the European Union. *Nature Energy, 2,* 17114.

Brunkard, J., Namulanda, G., & Ratard, R. (2008). Hurricane katrina deaths, louisiana, 2005. *Disaster Medicine and Public Health Preparedness, 2,* 215–223.

Carr, G. M., & Neary, J. P. (2008). Water quality for ecosystem and human health (2nd ed.) United Nations Environment Program Global Environment Monitoring System (GEMS)/Water Program.

DeFries, R. S., Rudel, T., Uriarte, M., & Hansen, M. (2010). Deforestation driven by urban population growth and agricultural trade in the twenty-first century. *Nature Geoscience, 3,* 178.

Duan, W., He, B., Chen, Y., Zou, S., Wang, Y., Nover, D., et al. (2018). Identification of long-term trends and seasonality in high-frequency water quality data from the Yangtze River basin, China. *Plos One, 13,* e188889.

Duan, W., Takara, K., He, B., Luo, P., Nover, D., & Yamashiki, Y. (2013). Spatial and temporal trends in estimates of nutrient and suspended sediment loads in the Ishikari River, Japan, 1985 to 2010. *Science of the Total Environment, 461,* 499–508.

Eliasson, J. (2015). The rising pressure of global water shortages. *Nature News, 517,* 6.

Fujii, Y., Satake, K., Sakai, S. I., Shinohara, M., & Kanazawa, T. (2011). Tsunami source of the 2011 off the Pacific coast of Tohoku Earthquake. *Earth, Planets and Space, 63,* 55.

Haraguchi, M., & Lall, U. (2015). Flood risks and impacts: A case study of Thailand's floods in 2011 and research questions for supply chain decision making. *International Journal of Disaster Risk Reduction, 14,* 256–272.

Horne, J., Tortajada, C., & Harrington, L. (2018). Achieving the Sustainable Development Goals: Improving water services in cities affected by extreme weather events. *International Journal of Water Resources Development, 34,* 475–489.

Konikow, L. F., & Kendy, E. (2005). Groundwater depletion: A global problem. *Hydrogeology Journal, 13,* 317–320.

Kundzewicz, Z. W., Mata, L. J., Arnell, N. W., Döll, P., Jimenez, B., Miller, K., et al. (2008). The implications of projected climate change for freshwater resources and their management. *Hydrological Sciences Journal, 53*(1), 3–10.

McCarthy, J. J. (2001). Climate change 2001: Impacts, adaptation, and vulnerability: Contribution of Working Group II to the third assessment report of the Intergovernmental Panel on Climate Change, Cambridge University Press.

Mekonnen, M. M., & Hoekstra, A. Y. (2016). Four billion people facing severe water scarcity. *Science Advances, 2,* e1500323.

Milly, P., Wetherald, R., Dunne, K. A., & Delworth, T. L. (2002). Increasing risk of great floods in a changing climate. *Nature, 415,* 514–517.

Oki, T., & Kanae, S. (2006). Global hydrological cycles and world water resources. *Science, 313,* 1068–1072.

Pachauri, R. K. (2008). Climate change 2007. In: Pachauri, R. K., & Reisinger, A. (Eds.) Synthesis Report. Contribution of Working Groups I, II and III to the Fourth Assessment Report. Geneva: IPCC.

Pearce, F. (2006). When the rivers run dry: water, the defining crisis of the twenty-first century, Beacon Press.

Polizzotto, M. L., Kocar, B. D., Benner, S. G., Sampson, M., & Fendorf, S. (2008). Near-surface wetland sediments as a source of arsenic release to ground water in Asia. *Nature, 454,* 505–508.

Harding, R., & Kabat P. (2007). Water and global change-The WATCH programme (pp. 50–51). Retrieved October, 2019, from http://www.eu-watch.org/media/default.aspx/emma/org/10001933/Water+and+Global+Change+%E2%80%93+the+WATCH+programme%5B1%5D.pdf.

Rosa, E. A., & Dietz, T. (2012). Human drivers of national greenhouse-gas emissions. *Nature Climate Change, 2,* 581.

Schade, C., & Pimentel, D. (2010). Population crash: prospects for famine in the twenty-first century. *Environment, Development and Sustainability, 12,* 245–262.

Vörösmarty, C. J., Green, P., Salisbury, J., & Lammers, R. B. (2000). Global water resources: vulnerability from climate change and population growth. *Science, 289,* 284.

Vörösmarty, C. J., McIntyre, P. B., Gessner, M. O., Dudgeon, D., Prusevich, A., Green, P., et al. (2010). Global threats to human water security and river biodiversity. *Nature, 467,* 555–561.

Wada, Y., van Beek, L. P., van Kempen, C. M., Reckman, J. W., Vasak, S., & Bierkens, M. F. (2010) Global depletion of groundwater resources. *Geophysical Research Letters, 37.*

Wu, Y., Ji, H., Wen, J., Wu, S., Xu, M., Tagle, F., et al. (2019). The characteristics of regional heavy precipitation events over eastern monsoon China during 1960–2013. *Global and Planetary Change, 172,* 414–427.

Zou, S., Jilili, A., Duan, W., Maeyer, P. D., & de Voorde, T. V. (2019). Human and natural impacts on the water resources in the Syr Darya River Basin. *Central Asia. Sustainability, 11,* 3084.

Chapter 2
Extreme Precipitation Events, Floods, and Associated Socio-Economic Damages in China in Recent Decades

2.1 Introduction

In recent decades, extreme precipitation events have increased in frequency and magnitude over the world under climate change, causing more water-related disasters such as floods, droughts, tropical cyclones, landslides, and tsunami (Amarnath et al. 2017; Dai 2013; Milly et al. 2002; Trenberth et al. 2014; Duan et al. 2016; Wu et al. 2019). There are lots of examples, such as Zhouqu mudflow disaster in 2010 (Wang 2013), Hurricane Katrina in 2005 (Pielke Jr et al. 2008), Amazon Drought in 2010 (Lewis et al. 2011), Heavy Rain Event of July 2018 in Japan (Tsuguti et al. 2019), Thai Floods in 2011 (Komori et al. 2012), and Kyushu's flash floods in 2014 (Duan et al. 2014). All these disasters have caused enormous loss of life and destruction and have therefore become a major obstacle to the realization of human security and sustainable socio-economic development (Jongman et al. 2014; Jonkman 2005; Smith and Katz 2013). The statistic obtained from the Global Emergency Disaster Database (EM-DAT) showed that there were 11,707 hydrological disasters from 1970 to 2013, killing more than 3,525,166 people, affecting more than 6.6 billion, and inflicting more than US$ 2,600 billion in damage. Also, these events caused the second disaster. For example, floods can generally cause serious environmental damage and pollution (Duan et al. 2013a, b). Therefore, it is very urgent and necessary to evaluate extreme precipitation events and water-related disasters.

China is vulnerable to extreme precipitation events (Sun et al. 2014). Some researchers have investigated and evaluated the changes of extreme climate events in China (Ji and Kang 2015; Tang et al. 2016; Zhang et al. 2006). Based on the gauged data and global climate model (GCM) outputs derived from the Coupled Model Intercomparison Project phases 3 and 5 (CMIP3 and CMIP5), most of these studies generally indicate that cold weather extremes exhibit a decreasing trend, warm weather extremes show an increasing trend, and extreme heavy precipitations are likely to become more severe (Chen et al. 2012; Li et al. 2017; Zhou et al. 2014).

Extreme precipitation events caused lots of floods in China. Generally, there were more than 1000 floods in the history of China, including the 1887 Yellow River floods

W. Duan and K. Takara, *Impacts of Climate and Human Activities on Water Resources and Quality*, https://doi.org/10.1007/978-981-13-9394-5_2

with a death toll of 0.9–2 million, the 1931 floods, with a death toll of 3–4 million, and the 1998 Yangtze River floods, with a death toll of 3704. In 2013, direct economic losses caused by floods were about 315.574 billion Yuan, occupying 0.55% of China's 2013 gross domestic product (GDP). As a result, research efforts have analyzed the causes of floods and contributed to the development of disaster management strategies (Du et al. 2015; Fischer et al. 2012; Zhang et al. 2009; Zhou et al. 2010). However, information of changes in floods and associated socio-economic damages in recent decades is still few.

This chapter is to evaluate the spatiotemporal distributions of extreme precipitation events, flooding trends, and socio-economic damages in China over recent decades. Specifically, multiple statistical methods are employed to describe the changes in flood trends and associated socio-economic damages. Also, possible causes such as extreme precipitation events are explored and discussed.

2.2 Data and Methods

2.2.1 Study Area and Data Sources

Located in Southeast Asia along the coastline of the Pacific Ocean, China covers an area of 9.6 million square kilometers, ranking the world's third largest country. The land topography can be divided into three levels, which causes that most rivers in China flow from west to east (Fig. 2.1) and most water-related disasters appear in the eastern plains. We obtained the water-related disaster database from the Bulletin of Flood and Drought Disasters in China (2006–2013), which is annually edited from 2006 by the State Flood Control and Drought Relief Headquarters and the Ministry of Water Resources, China. Flood and drought statistics from local to central authorities are recorded in the bulletin, which mainly contain location, time and date, flooded river, flooded villages, towns or settlements, death toll, losses, and other associated socio-economic damages. A flood is defined as waterlogging caused by heavy rainfall, typhoons, rainstorms, dam or riverbank collapse, snow melting, flash floods, debris flows, mountain torrents, etc.

Information about the flood disaster was collected from the EM-DAT International disasters database, flood risk studies and reports, and the gross domestic product (GDP) for 31 provinces in China were obtained from the National Bureau of Statistics of China (www.stats.gov.cn). Moreover, the observed daily precipitation data from 1961 to 2010 with $0.5^\circ \times 0.5^\circ$ resolution were selected to compute the extreme precipitation for China, which could be used to estimate the frequency of the floods. This dataset was charged by the National Climate Center, China Meteorological Administration, which was developed from 2416 weather monitoring stations (Wu and Gao 2013).

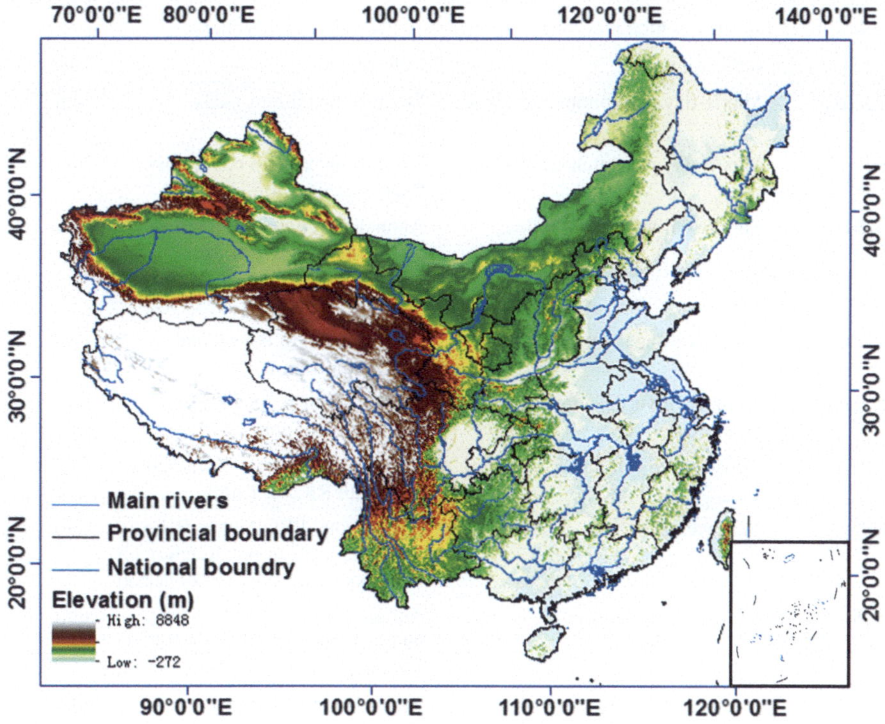

Fig. 2.1 Location, elevations, provinces, and main rivers of China

2.2.2 *Methods*

2.2.2.1 Computation of Extreme Precipitation Indices

A total of 27 widely used climate indices, which were developed and recommended by the Expert Team on Climate Change Detection and Indices (ETCCDI) (available at http://www.climdex.org/indices.html) for describing and assessing climate extreme events (Duan et al. 2015; Sillmann et al. 2013; Zhang et al. 2011). Most floods are due to excessive precipitation, so only nine climate indices (see Table 2.1) including the annual total wet-day precipitation (PRCPTOT), the annual maximum 1-day precipitation (RX1day), the annual maximum 5-day precipitation (RX5day), the annual number of heavy precipitation days (R10mm), the annual number of very heavy precipitation days (R20mm), the simple daily intensity index (SDII), the annual total precipitation on very wet days (R95p), the annual total precipitation on extremely wet days (R99p), and the consecutive wet days (CWD) were therefore computed to evaluate changes of extreme precipitation.

Table 2.1 Definitions of nine precipitation extreme indices used in this chapter

ID	Indicator name	Definitions	Units
RX1day	Max 1-day precipitation amount	Monthly maximum 1-day precipitation	mm
RX5day	Max 5-day precipitation amount	Monthly maximum consecutive 5-day precipitation	mm
SDII	Simple daily intensity index	Annual total precipitation divided by the number of wet days (defined as PRCP >= 1.0 mm) in the year	mm/day
R10mm	Number of heavy precipitation days	Annual count of days when PRCP >= 10 mm	day
R20mm	Number of very heavy precipitation days	Annual count of days when PRCP >= 20 mm	day
CWD	Consecutive wet days	Maximum number of consecutive days with RR >= 1 mm	day
R95p	Very wet days	Annual total PRCP when RR > 95th percentile of precipitation on wet days in the 1961–1990 period	mm
R99p	Extremely wet days	Annual total PRCP when RR > 99th percentile of precipitation on wet days in the 1961–1990 period	mm
PRCPTOT	Annual total wet-day precipitation	Annual total PRCP in wet days (RR >= 1 mm)	mm

Abbreviations are as follows: RR, daily precipitation. A wet day is defined when RR >= 1 mm, and a dry day when RR < 1 mm

2.2.2.2 Trend Analysis of Disasters

After the extraction process, disaster data (e.g., the number of incidents, etc.) are categorized and transformed into an Excel database. Trend analysis was completed using the Kendall's Tau test for monotonic trends to determine if statistically significant trends exist in floods and associated socio-economic damages through time. Mann–Kendall test is a non-parametric rank-based statistical test (Kendall 1975; Mann 1945). The Mann–Kendall trend test can be stated most generally as a test for whether Y values tend to increase or decrease with T (monotonic change). Tau values are considered statistically significant at $p \leq 0.05$.

The Mann–Kendall S statistic is computed as follows:

$$S = \sum_{i=1}^{n-1} \sum_{j=i+1}^{n} \text{sign}(T_j - T_i) \tag{2.1}$$

$$\text{sign}(T_j - T_i) = \begin{cases} 1 & if \quad T_j - T_i > 0 \\ 0 & if \quad T_j - T_i = 0 \\ -1 & if \quad T_j - T_i < 0 \end{cases} \tag{2.2}$$

where T_j and T_i are the water-related disaster variabilities at multiple time scales j and i, $j > i$, respectively. When $n \geq 10$, the statistic S is approximately normally distributed with the mean and variance as follows:

$$E(S) = 0 \tag{2.3}$$

The variance (σ^2) for the S—statistic is defined by

$$\sigma^2 = \frac{n(n-1)(2n+5) - \sum t_i(i)(i-1)(2i+5)}{18} \tag{2.4}$$

where t_i denotes the number of ties to extent i. The summation term in the numerator is used only if the data series contains tied values. The standard test statistic Z_S is calculated as follows:

$$Z_S = \begin{cases} \frac{S-1}{\sigma} & for\ S > 0 \\ 0 & for\ S = 0 \\ \frac{S+1}{\sigma} & for\ S < 0 \end{cases} \tag{2.5}$$

The statistic test (Z_S) is used as a measure of significance of trend (e.g., Yue et al. 2002). This test statistic is used to test the null hypothesis (H_0). If $|Z_S|$ is greater than $Z_{\alpha/2}$, where α is the chosen significance level (e.g., 5% with $Z_{0.025} = 1.96$) then the null hypothesis is invalid implying that the trend is significant.

2.3 Results

2.3.1 Changes of Extreme Precipitation Events

Figure 2.2 shows the spatial distribution of mean annual precipitation extremes indices from 1961 to 2010, suggesting an uneven spatial distribution for all nine indices. Generally, the southeast region (e.g., Yangtze River Basin and Pearl River Basin) had higher values for all indices than the northwest region (e.g., Tarim River Basin). Precipitation was mainly concentrated in Guangdong, Fujian, Jiangxi, Hunan, and Zhejiang Provinces, the mean annual value of which was up to 2000 mm during the period 1961–2010; however, the mean regional annual precipitation was less than 100 mm in Xinjiang Province. The mean annual value of R10m was ranging from 0 day in northwest region to 80 days in southeast region, which means that there were more days with precipitation ≥10 mm in southeast region. The mean annual

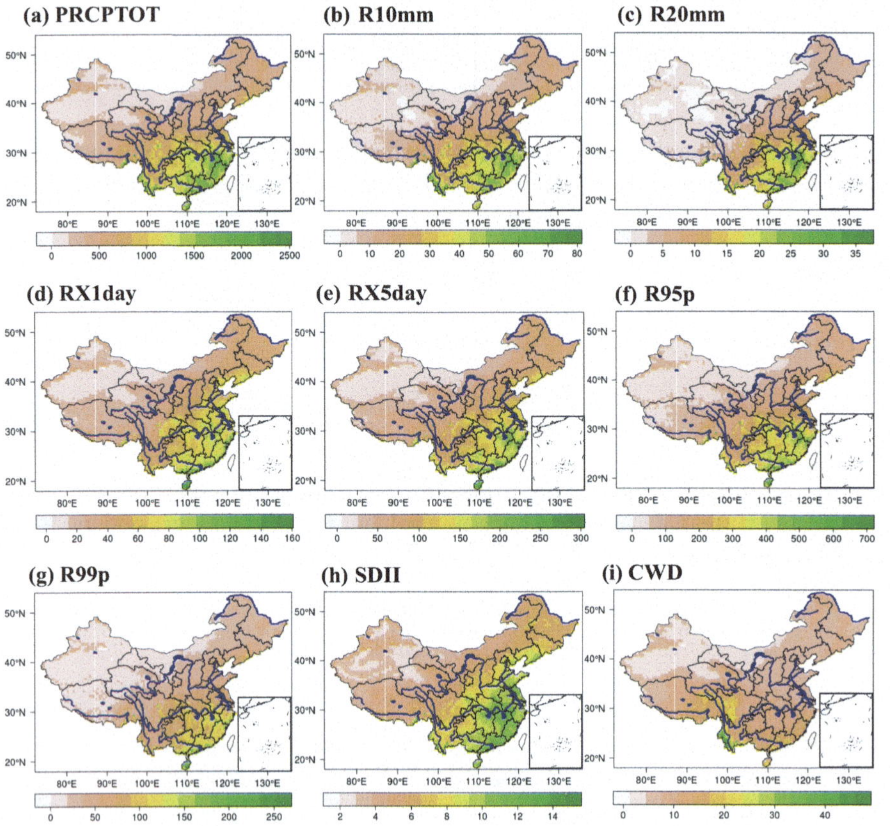

Fig. 2.2 Spatial pattern of mean annual precipitation extreme indices from 1961 to 2010

values of RX1day were greater than 100 mm in many regions of the Yangtze River Basin and Pearl River Basin, possibly leading to more landslides and flash floods.

As shown in Fig. 2.3, an increasing trend was found for most indices in most regions over China. The largest trend of precipitation was found in the southeast corner of Xizang Province, up to 150 mm/10a. The middle reaches of the Yangtze River had the higher trend of R20mm (up to 3 days/10a) compared with other regions, revealing an upward trend in rainy days with daily rainfall $\geq$ 20 mm; moreover, same distributions of higher values were found for RX1day, RX5day, R95p, R99p, and SDII in the middle reaches of the Yangtze River, which are the flood-prone areas in China. The grid with an increasing trend at the 95% confidence level for PRCPTOT, R10m, and R20m tended to be clustered in Xinjiang, Xizang, Qinghai, and Gansu Provinces. For consecutive wet days (CWD), as shown in Fig. 2.2i, a decreasing trend was found in most areas of Yunnan Province, suggesting that the number of consecutive dry days increased over the 50 years.

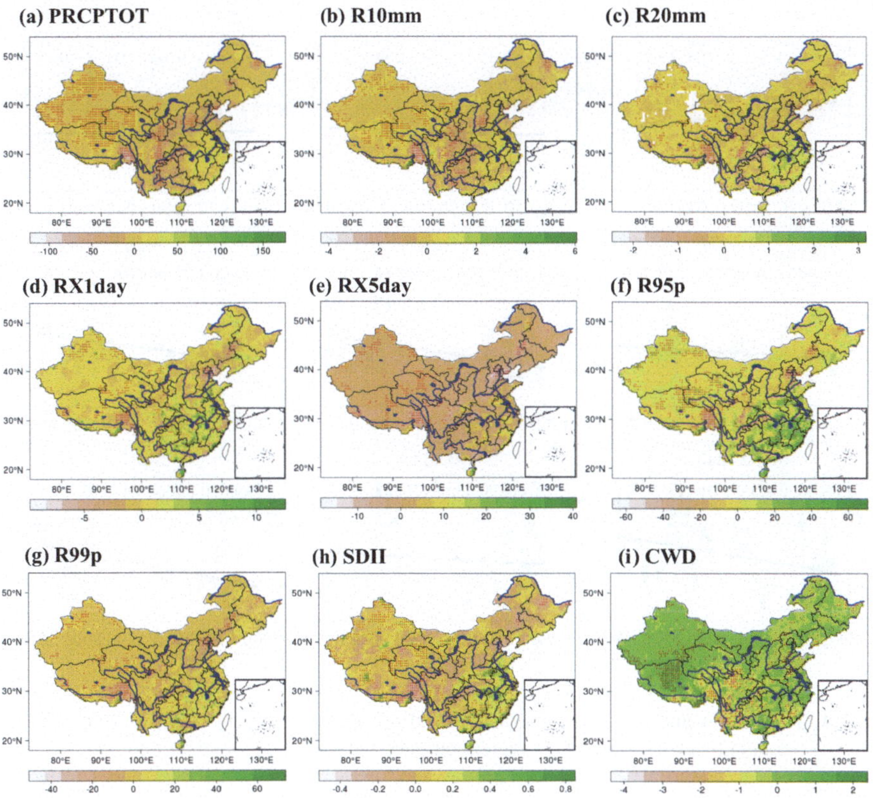

Fig. 2.3 Spatial pattern of trends for annual precipitation extreme indices from 1961 to 2010. Trends that are significant at the 95% level are circled

Figure 2.4 illustrates time series and trends for regional annual precipitation extreme indices from 1961 to 2010, which suggests that CWD and R10mm had a downward trend, while an upward trend for all other precipitation extreme indices. Moreover, five indices (CWD, R95p, R99p, RX1day, and SDII) had statistically significant trends. Of them, CWD had a decreasing trend at the 95% confidence level, the value of which was up to 2.69 days/10a; the other four indices including R95p (Z = 2.36), R99p (Z = 2.04), RX1day (Z = 2.19), and SDII (Z = 2.98) had an increasing trend at the 95% confidence level, the trend values of which were 3.12 mm/10a, 1.28 mm/10a, 0.42 mm/10a, and 0.06 mm/10a, respectively. This phenomenon indicates that although the number of consecutive wet days decreased, the extreme precipitation such as max 1-day precipitation amount increased from 1961 to 2010 over China, which probably caused more extreme events or floods.

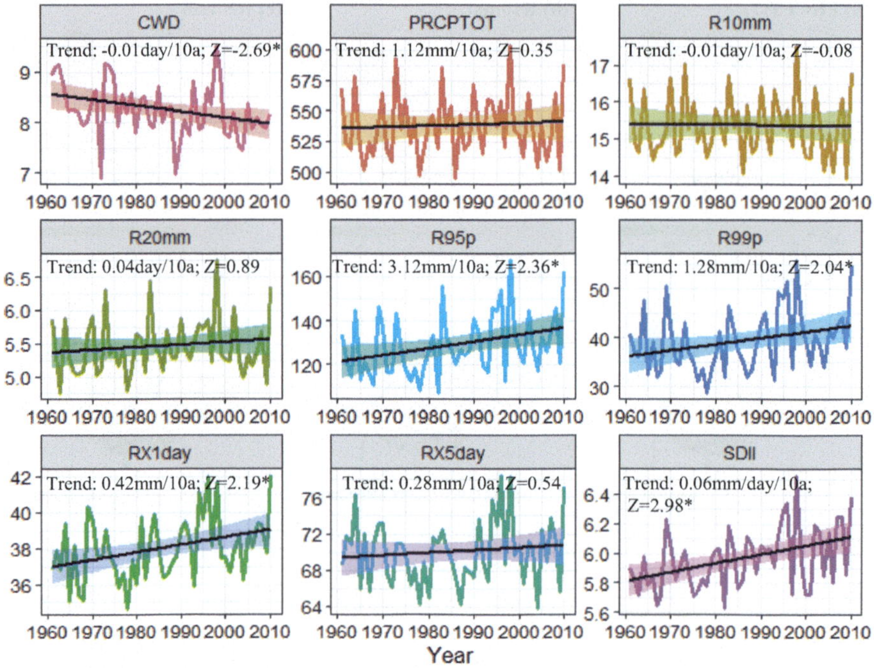

Fig. 2.4 Time series and trends for regional annual precipitation extreme indices from 1961 to 2010. The straight black lines are linear regression lines and the shades are 95% confidence band

2.3.2 Changes of Flooding Frequency

As shown in Fig. 2.5, the proportions of more serious 20-year floods, 10–20-year floods, and 5–10-year floods were 31 (14.6%), 55 (25.8%), and 127 (59.6%), respectively, suggesting that 5–10-year flood was the main problem in flood disasters in China. Also, 58 floods occurred in the Yangtze River Basin, 27.2% of all floods, followed by the Huaihe River Basin (27, 12.7%), the Pearl River Basin (26, 12.2%), and the Liaohe River Basin (26, 12.2%). In terms of the number of more serious 20-year floods, the Yangtze River Basin had the most frequent floods, with 6, followed by the Pearl River Basin (5), the Yellow River Basin (4), and the Huaihe River Basin (4), showing that catastrophic floods were generally concentrated on the plains of the Yangtze River, the Pearl River, the Yellow River, and the Huaihe River. All these distributions are in line with the precipitation amounts (see Fig. 2.2).

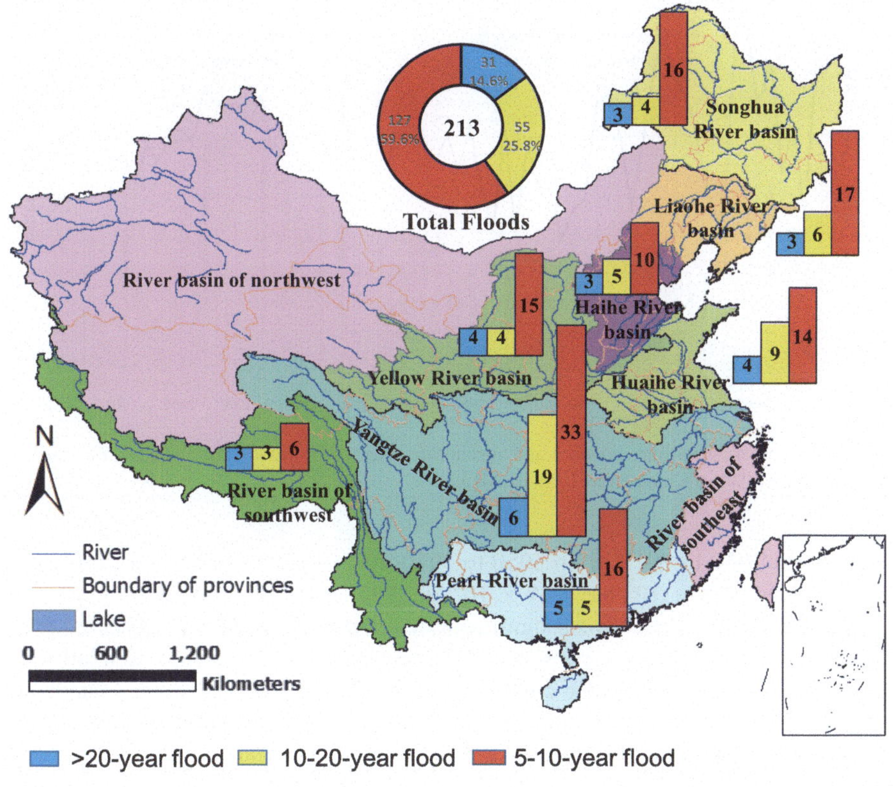

Fig. 2.5 Distribution of three levels of floods including 20-, 10-, and 5-year floods in major rivers of China in twentieth century

2.3.3 *Impacts on Agriculture*

Table 2.2 shows results of the mean value and the Mann–Kendall test for associated socio-economic damages. Figure 2.6 shows the nationwide trend in China. The flood-covered agricultural area has generally increased from 1950 to 2013. The average

Table 2.2 Results of the mean value and the Mann–Kendall test for associated socio-economic damages

Name	Mean	Z_value
Covered area	9.38×10^6 ha	4.04
Affected area	5.44×10^6 ha	3.25
Death tolls	4387 people	−4.04
Death tolls from mountain torrents	–	−2.14
Destroyed houses	189.94×10^4 houses	0.38
Direct economic damage	1.38×10^{12} Yuan	2.46

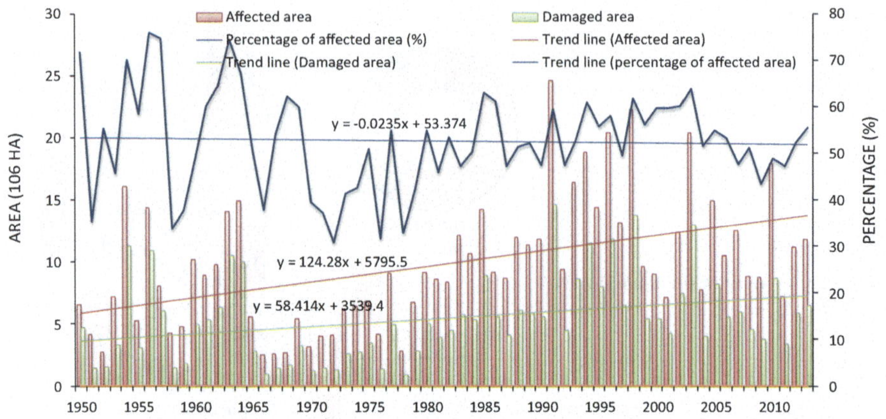

Fig. 2.6 Areas of agriculture covered and agriculture affected by floods (10^6 ha), and percentage of affected area (%) in China from 1950 to 2013

values from 1950 to 2013, from 1950 to 1970, from 1971 to 1990, and from 1991 to 2013 were 9.38×10^6 ha, 7.04×10^6 ha, 8.32×10^6 ha, and 13.46×10^6 ha, respectively. Figure 2.3 shows a significant increase in coverage area observed over the past 64 years, which is consistent with the results of the Mann–Kendall test, which is significant at the 95% confidence level ($Z = 4.04$). The mean value from 1991 to 2013 increased by two orders of magnitude from the average of 1950–1970, with the largest loss in 1991, reaching 24.60×10^6 ha hectares, followed by 1998 (22.29×10^6 ha).

As with the coverage area, the area of agriculture affected by floods over the past 64 years was also on the rise (the linear trend line in Fig. 2.6). The maximum was found in 1991, reaching 14.61×10^6 ha, followed by 1998 (13.79×10^6 ha). However, the proportion of affected areas has declined slightly, probably because of increased agricultural areas and improved mitigation measures. The China Agricultural Statistical Yearbook 2014 showed an increase of 14.62% from 1985 to 2013. Changes in coverage and affected areas indicated frequent flooding in China, especially in recent decades.

Figure 2.7 shows that there was a steady decline in flood-related deaths from 1950 to 2013. Average flood-related deaths have dropped by nearly two-thirds from 6146 in the time period of 1950–1970. The results show a significant decrease in deaths, consistent with the results of the Mann–Kendall test, with a 95% confidence level ($Z = -4.04$). As can be seen from Fig. 2.7, the most deadly flood occurred in 1954, with a maximum of 42,447 deaths, followed by 1975 (29,653 deaths). From June to September 1954, a series of catastrophic floods occurred in Hubei Province in the Yangtze River Basin. Due to the unusually heavy precipitation and the extraordinarily long rainy season in the middle reaches of the Yangtze River in the late spring of 1954, about 33,000 people died, including those who died of the plague after the disaster. In August 1975, a typhoon caused two large dams (Boqiao Dam and Shimendan

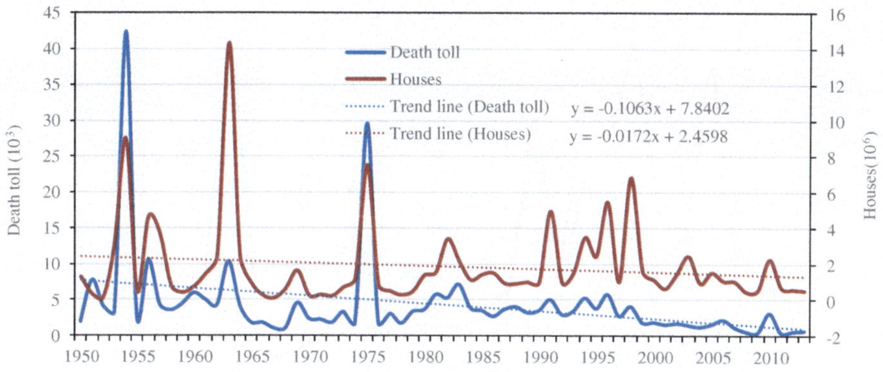

Fig. 2.7 Number of deaths from floods and houses destroyed by floods in China from 1950 to 2013

Dam) to fail in Henan Province, China (Xu et al. 2008), resulting in approximately 26,000 deaths from floods and another 145,000 due to the subsequent epidemics and famine.

At the same time, mountain torrents disasters have increased due to the increasing frequency of extreme precipitation events (see Fig. 2.4). Figure 2.8 shows the number of deaths caused by mountain torrents from 1991 to 2013 and the percentage change in floods in China. Judging from the trend of flood deaths, the number of deaths caused by mountain torrents from 1991 to 2013 showed a downward trend. Due to the 2010 Zhouqu mudslide disaster, the number of deaths was the highest in 2010 (Xiao et al. 2013). However, the proportion of deaths caused by floods exhibited an increasing trend, indicating that mountain torrents have become a major threat to the security of people and properties in recent years.

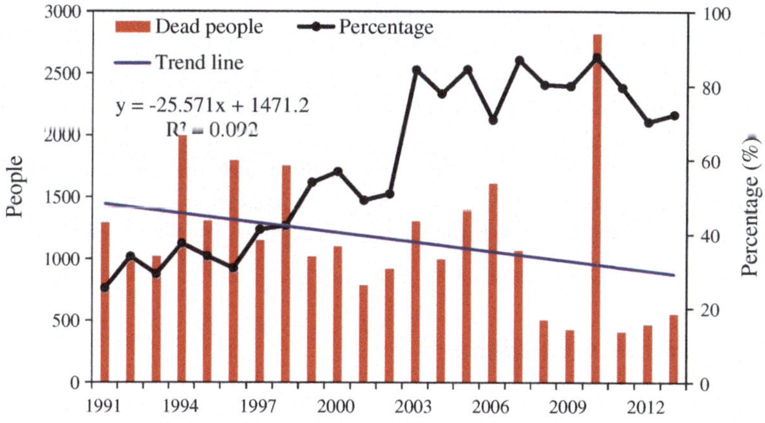

Fig. 2.8 Number of deaths from mountain torrents disaster and proportion of floods in China from 1990 to 2013

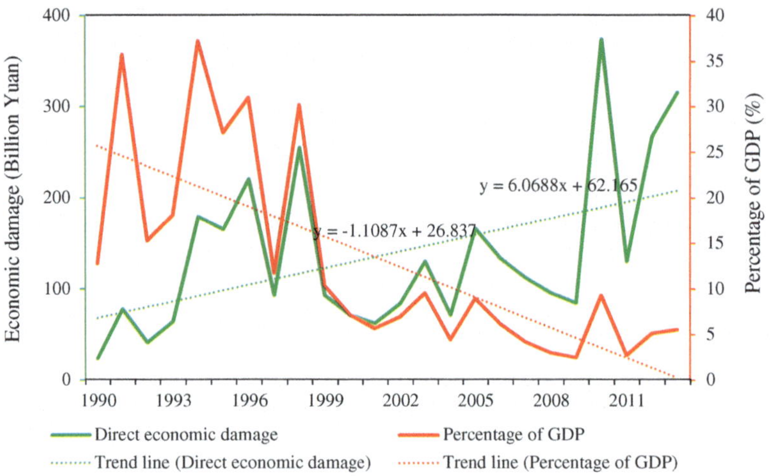

Fig. 2.9 Annual direct economic damage and proportion of GDP in China from 1990 to 2013

Figure 2.8 also shows that the number of houses destroyed by floods has also decreased steadily from 1950 to 2013, but in smaller magnitude. The largest number of destroyed houses was observed in 1963, up to 14.4×10^6, followed by 1954 (9.009×10^6). In 1963, a serious flood occurred in the Haihe River, which caused dam failure to 330 small-sized reservoirs, 2,400 breaches in levees, destroyed 62% of the irrigation area, and submerged 90% of drainage engineering infrastructure.

Figure 2.9 shows the changes in direct economic losses and GDP ratios from 1990 to 2013. Direct economic losses had increased significantly over the past 24 years, but the GDP ratio had exhibited a decreasing trend, especially from 1999. The reason for this phenomena was possibly due to China's torrid growth over the past decade. Both results were supported by the Mann–Kendall trend analysis. The direct economic losses that occurred in 2010 were the highest, reaching 374.5 billion yuan, accounting for 9.33% of GDP. The 2010 Chinese flood began in the Yangtze River Basin in early May 2010. Due to unusual climate patterns including an El Niño "Modoki" and the jet stream, the damage caused by these floods had exceeded the level of the 1998 Yangtze River floods. Heavy rains occurred in most parts of southwestern China, causing large-scale floods in July and August 2013, causing the second largest direct economic loss, with up to 315.6 billion yuan, accounting for 5.55% of GDP.

2.3.4 Distribution Variations

Figure 2.10 shows the provincial distribution of the average value of affected people, mortality, direct economic damages, crop covered areas, crop affected areas, and crop areas destroyed by floods from 2006 to 2013. Results reveal that floods mainly

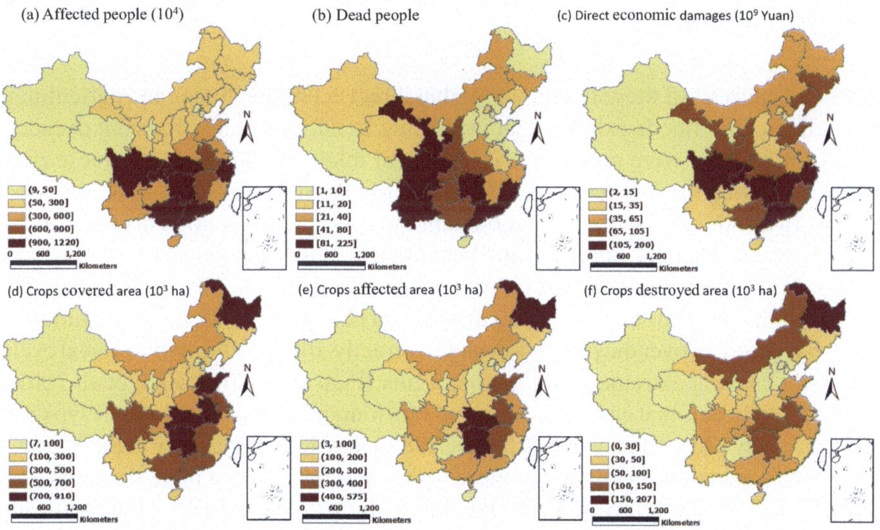

Fig. 2.10 Distribution of floods across China expressed on the basis of provinces and expressed as average value of **a** affected people, **b** dead people, **c** direct economic damages, **d** crop covered areas, **e** crop affected areas, and **f** crop destroyed areas from 2006 to 2013

occurred in the Yangtze River Basin, especially in Sichuan, Chongqing, Hunan, Jiangxi, and Shanghai. These areas have a common feature that they are prone to extreme rainfall, which have been illustrated in Fig. 2.3. In addition, Fig. 2.10b shows a large number of deaths in Gansu Province in 2010 due to destructive debris flows. Specifically, the heavy rains triggered catastrophic debris flows in the catchments of the Sanyanyu and Luojiayu torrents, Zhouqu County, Gansu Province, on August 7, 2010, which killed 1,765 people (Tang et al. 2011). Also, Fig. 2.10a–c indicates that typhoons appear to be an important driver of annual deaths and huge economic losses, especially in the southeastern coastal regions of China, including Guangxi, Guangdong, and Fujian.

Figure 2.10e–f suggests that serious destruction of agricultural land was observed in Heilongjiang Province, which was the same with Hunan, Hubei, and Anhui Provinces. This phenomena was mainly due to the 2013 China–Russia floods. In mid-August 2013, severe flooding occurred in eastern Russia and parts of northeastern China (including Heilongjiang, Jilin, and Liaoning Provinces), killing at least 85 people, destroying more than 60,000 homes, and destroying more than 787,000 ha of agricultural land in the heavily agricultural region.

2.4 Discussions

Statistical analysis of these floods shows that direct economic losses and agricultural areas affected by floods exhibited an increasing trend, especially in recent decades. Most of the floods were concentrated in several relatively developed Chinese river basins, including the Yangtze River Basin and the Pearl River Basin. In addition, an increasing trend was also found in mountain torrents and urban flood disasters in the past few years. These features are attributed to the following main factors.

First of all, the increase in extreme precipitation events caused by climate change is an important driver for these flooding disasters. The regional and seasonal rainfall in China has changed over the past few decades, directly affecting the hydrological cycle and easily causing extreme precipitation events (Piao et al. 2010; Zhai et al. 2005). Figure 2.3 indicates that an increasing trend was found in R95p (Z = 2.36), R99p (Z = 2.04), RX1day (Z = 2.19), and SDII (Z = 2.98) over China, especially in the Yangtze River Basin, which probably led to more floods and thus caused more damages (see Fig. 2.10). As shown in Fig. 2.11, the July–August extreme precipitation events exhibited an increasing trend in China, especially in the southeastern regions, which reveals there will be more floods in the Yangtze River Basin in summer. For example, Song et al. (2011) found that the number of precipitation days with precipitation or rainfall amount above 50 mm had an increasing trend in most of southeastern China, with important implications for people's lives and socio-economic development. Jiang et al. (2008) and Su et al. (2006) argued that extreme precipitation events tended to exacerbate flood disasters in the middle and lower reaches of the Yangtze River in summer. Also, extreme precipitation is the main driver of mudslides in China's central and southern regions due to moderate-to-strong soil erosion risks in these regions, including Shaanxi, Gansu, Yunnan, and Sichuan. These areas also have significant risks of mudslides after heavy rainfall. The major risk areas of mudslides in China account for about 13% of the country's total area (Liu et al. 2011; Liu

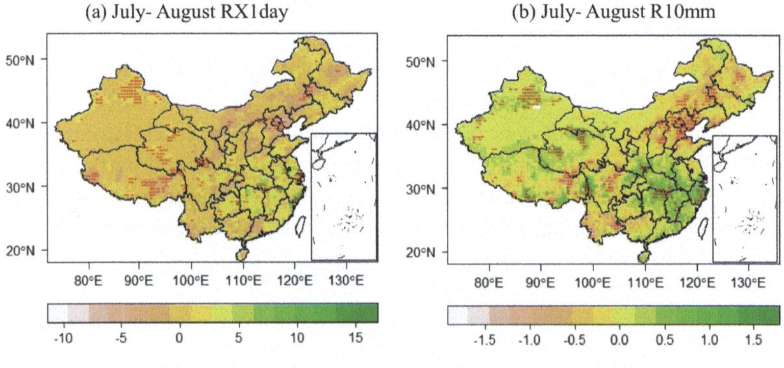

Fig. 2.11 Spatial distribution of the trend of **a** the July–August RX1day (mm per decade) and **b** the July–August R10mm (days per year) from 1961 to 2010, and areas with red dots indicate 95% significance

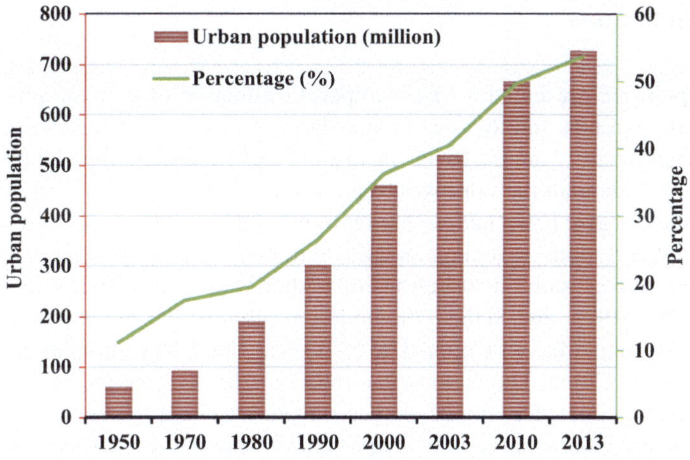

Fig. 2.12 Time series of urbanization in China from 1950 to 2013

and Ni 2005) and the Zhouqu mudslide disaster is a typical mudslide caused by heavy rains (Xiao et al. 2013). Considering the increase in future climate extremes (Sun et al. 2015), mudslide disasters will likely to cause serious damage to humans, buildings, and the natural environment. Therefore, governments especially in these areas at significant risk of mudflow occurrence should implement rainfall-monitoring programs to provide the early warning system for defending potential debris flows.

In addition, urbanization pressures caused by population growth and rapid economic development have put the number of individuals and properties at risk. As shown in Fig. 2.12, China only spent 22 years in urbanization rate from 20 to 40%, which was much faster than France (100 years), Germany (80 years), and USA (40 years). In 2013, China's urbanization rate reached 53.73%, and there were 86 cities with a population over 1 million. Such a rapid urbanization has two main negative effects on urban flooding. First, during this rapid urbanization process, the increasing urban impermeable surfaces may reduce natural feeding of underground aquifers (Luo et al. 2015; Parsasyrat and Jamali 2015). Also, rapid urbanization has significantly changed the geomorphological complexity of the urban river networks, and low-grade rivers have been severely damaged in the process of urbanization in China (Xu et al. 2010; Zhang et al. 2015). These two main factors have a negative impact on the urban hydrological processes, such as accelerating runoff flow velocity and enlarging peak flow (Grove et al. 2001; Li and Wang 2009), thereby increasing the urban flood risks. Urban flood risk had been on the rise during recent decades (Chen et al. 2015; Zheng et al. 2013), and statistical analysis shows that up to 62% of the 351 cities surveyed suffered severe floods between 2008 and 2010, and 137 cities of them had been hit more than 3 times. Besides, due to high-density buildings and underground infrastructures (such as subways and many kinds of pipelines in China), it is difficult to build and improve the flood infrastructure (Chen et al. 2015), which may increase the risk of flash floods in the future.

2.5 Conclusions

In this chapter, we presented a spatiotemporal evaluation of extreme precipitation events, flooding trends, and socio-economic damages in China. In recent decades, the southeast region (e.g., Yangtze River Basin and Pearl River Basin) had higher values for all nine indices than the northwest region (e.g., Tarim River Basin); the middle reaches of the Yangtze River had the higher trend of R20mm (up to 3 days/10a) compared with other regions, revealing an upward trend in rainy days with daily rainfall $\geq$20 mm; moreover, same distributions of higher values were found for RX1day, RX5day, R95p, R99p, and SDII in the middle reaches of the Yangtze River, which are the flood-prone areas in China. The 5–10-year flood was the main problem in flood disasters in China. The nationwide trend for China is that the area of agriculture covered and affected by floods increased from 1950 to 2013, with an average of area covered and affected area over the period 1991 to 2013 of 13.46×10^6 ha and 7.45×10^6 ha, respectively, while a significant downtrend in death tolls exhibited a significant downtrend, with the deadliest floods in 1954 (42,447 deaths). In recent years, direct economic losses have increased, and the loss in 2010 was the largest, up to 374.5 billion Yuan, accounting for 9.33% of the GDP. The damage caused by the floods was mainly concentrated on the plains along the Yangtze River, the Pearl River, the Yellow River, and other large rivers, which will become more vulnerable due to climate changes and rapid urbanization.

References

Amarnath, G., Yoshimoto, S., Goto, O., Fujihara, M., Smakhtin, V., Aggarwal, P.K., et al. (2017). Global trends in water-related disasters using publicly available database for hazard and risk assessment. Retrieved August, 2019, from https://cgspace.cgiar.org/bitstream/handle/10568/93032/H048407.pdf.

Chen, H., Sun, J., Chen, X., & Zhou, W. (2012). CGCM projections of heavy rainfall events in China. *International Journal of Climatology, 32,* 441–450.

Chen, Y., Zhou, H., Zhang, H., Du, G., & Zhou, J. (2015). Urban flood risk warning under rapid urbanization. *Environmental Research, 139,* 3–10.

Dai, A. (2013). Increasing drought under global warming in observations and models. *Nature Climate Change, 3,* 52–58.

Du, X., Jin, X., Yang, X., Yang, X., Xiang, X., & Zhou, Y. (2015). Spatial-temporal pattern changes of main agriculture natural disasters in China during 1990–2011. *Journal of Geographical Sciences, 25,* 387–398.

Duan, W., He, B., Nover, D., Fan, J., Yang, G., Chen, W., et al. (2016). Floods and associated socioeconomic damages in China over the last century. *Natural Hazards, 82,* 401–413.

Duan, W., He, B., Takara, K., Luo, P., Hu, M., Alias, N. E., et al. (2015). Changes of precipitation amounts and extremes over Japan between 1901 and 2012 and their connection to climate indices. *Climate Dynamics, 45,* 2273–2292.

Duan, W., He, B., Takara, K., Luo, P., Nover, D., Sahu, N., et al. (2013a). Spatiotemporal evaluation of water quality incidents in Japan between 1996 and 2007. *Chemosphere, 93,* 946–953.

Duan, W., He, B., Takara, K., Luo, P., Nover, D., Yamashiki, Y., et al. (2014). Anomalous atmospheric events leading to Kyushu's flash floods, July 11–14, 2012. *Natural Hazards, 73,* 1255–1267.

Duan, W., Takara, K., He, B., Luo, P., Nover, D., & Yamashiki, Y. (2013b). Spatial and temporal trends in estimates of nutrient and suspended sediment loads in the Ishikari River, Japan, 1985 to 2010. *Science of the Total Environment, 461,* 499–508.

Fischer, T., Gemmer, M., Liu, L., & Su, B. (2012). Change-points in climate extremes in the Zhujiang River Basin, South China, 1961–2007. *Climatic Change, 110,* 783–799.

Grove, M., Harbor, J., Engel, B., & Muthukrishnan, S. (2001). Impacts of urbanization on surface hydrology, Little Eagle Creek, Indiana, and analysis of LTHIA model sensitivity to data resolution. *Physical Geography, 22,* 135–153.

Ji, Z., & Kang, S. (2015). Evaluation of extreme climate events using a regional climate model for China. *International Journal of Climatology, 35,* 888–902.

Jiang, T., Kundzewicz, Z. W., & Su, B. (2008). Changes in monthly precipitation and flood hazard in the Yangtze River Basin, China. *International Journal of Climatology, 28,* 1471–1481.

Jongman, B., Hochrainer-Stigler, S., Feyen, L., Aerts, J. C., Mechler, R., Botzen, W. W., et al. (2014). Increasing stress on disaster-risk finance due to large floods. *Nature Climate Change, 4,* 264–268.

Jonkman, S. N. (2005). Global perspectives on loss of human life caused by floods. *Natural Hazards, 34,* 151–175.

Kendall, M.G. (1975). *Rank correlation methods: Charles Griffin*. London.

Komori, D., Nakamura, S., Kiguchi, M., Nishijima, A., Yamazaki, D., Suzuki, S., et al. (2012). Characteristics of the 2011 Chao Phraya River flood in central Thailand. *Hydrological Research Letters, 6,* 41–46.

Lewis, S. L., Brando, P. M., Phillips, O. L., van der Heijden, G. M., & Nepstad, D. (2011). The 2010 amazon drought. *Science, 331,* 554.

Li, Y., & Wang, C. (2009). Impacts of urbanization on surface runoff of the Dardenne Creek watershed, St. Charles County, Missouri. *Physical Geography, 30,* 556–573.

Li, Z., Xu, X., Liu, M., Li, X., Zhang, R., Wang, K., et al. (2017). State-space prediction of spring discharge in a karst catchment in southwest China. *Journal of Hydrology, 549,* 264–276.

Liu, R., & Ni, J. (2005). Landslide and rock fall hazard zonation in China. *Journal of Basic Science and Engineering, 13,* 13–22. (in Chinese with English abstract).

Liu, X. L., Yu, C. J., & Shang, Z. H. (2011). Risk mapping and spatial pattern of debris flow and landslide hazards in China. *Journal of Basic Science and Engineering, 19,* 721–731. (in Chinese with English abstract).

Luo, P., He, B., Duan, W., Takara, K., & Nover, D. (2015). Impact assessment of rainfall scenarios and land-use change on hydrologic response using synthetic Area IDF curves. *Journal of Flood Risk Management.*

Mann, H. B. (1945). Nonparametric tests against trend. *Econometrica, 13,* 245–259.

Milly, P., Wetherald, R., Dunne, K. A., & Delworth, T. L. (2002). Increasing risk of great floods in a changing climate. *Nature, 415,* 514–517.

Parsasyrat, L., & Jamali, A. A. (2015). the effects of impermeable surfaces on the flooding possibility in Zarrin-Shahr, Isfahan Municipal Watershed. *Journal of Applied Environmental and Biological Sciences, 5,* 28–38.

Piao, S., Ciais, P., Huang, Y., Shen, Z., Peng, S., Li, J., et al. (2010). The impacts of climate change on water resources and agriculture in China. *Nature, 467,* 43–51.

Pielke, R. A., Jr., Gratz, J., Landsea, C. W., Collins, D., Saunders, M. A., & Musulin, R. (2008). Normalized hurricane damage in the United States: 1900–2005. *Natural Hazards Review, 9,* 29–42.

Sillmann, J., Kharin, V. V., Zhang, X., Zwiers, F. W., & Bronaugh, D. (2013). Climate extremes indices in the CMIP5 multimodel ensemble: Part 1. Model evaluation in the present climate. *Journal of Geophysical Research: Atmospheres, 118,* 1716–1733.

Smith, A. B., & Katz, R. W. (2013). US billion-dollar weather and climate disasters: Data sources, trends, accuracy and biases. *Natural Hazards, 67,* 387–410.

Song, Y., Achberger, C., & Linderholm, H. W. (2011). Rain-season trends in precipitation and their effect in different climate regions of China during 1961–2008. *Environmental Research Letters, 6,* 34025.

Su, B. D., Jiang, T., & Jin, W. B. (2006). Recent trends in observed temperature and precipitation extremes in the Yangtze River basin, China. *Theoretical and Applied Climatology, 83,* 139–151.

Sun, Q., Miao, C., & Duan, Q. (2015). Extreme climate events and agricultural climate indices in China: CMIP5 model evaluation and projections. *International Journal of Climatology.*

Sun, Y., Zhang, X., Zwiers, F. W., Song, L., Wan, H., Hu, T., et al. (2014). Rapid increase in the risk of extreme summer heat in Eastern China. *Nature Climate Change, 4,* 1082–1085.

Tang, C., Rengers, N., van Asch, T. W., Yang, Y. H., & Wang, G. F. (2011). Triggering conditions and depositional characteristics of a disastrous debris flow event in Zhouqu city, Gansu Province, northwestern China. *Natural Hazards and Earth System Science, 11,* 2903–2912.

Tang, J., Niu, X., Wang, S., Gao, H., Wang, X., & Wu, J. (2016). Statistical downscaling and dynamical downscaling of regional climate in China: Present climate evaluations and future climate projections. *Journal of Geophysical Research: Atmospheres, 121,* 2110–2129.

Trenberth, K. E., Dai, A., van der Schrier, G., Jones, P. D., Barichivich, J., Briffa, K. R., et al. (2014). Global warming and changes in drought. *Nature Climate Change, 4,* 17–22.

Tsuguti, H., Seino, N., Kawase, H., Imada, Y., Nakaegawa, T., & Takayabu, I. (2019). Meteorological overview and mesoscale characteristics of the Heavy Rain Event of July 2018 in Japan. *Landslides, 16,* 363–371.

Wang, G. L. (2013). Lessons learned from protective measures associated with the 2010 Zhouqu debris flow disaster in China. *Natural Hazards, 69,* 1835–1847.

Wu, J., & Gao, X. (2013). A gridded daily observation dataset over China region and comparison with the other datasets. *Chinese Journal of Geophysics, 56,* 1102–1111.

Wu, Y., Ji, H., Wen, J., Wu, S., Xu, M., Tagle, F., et al. (2019). The characteristics of regional heavy precipitation events over eastern monsoon China during 1960–2013. *Global and Planetary Change, 172,* 414–427.

Xiao, H., Luo, Z., Niu, Q., & Chang, J. (2013). The 2010 Zhouqu mudflow disaster: Possible causes, human contributions, and lessons learned. *Natural Hazards, 67,* 611–625.

Xu Youpeng, X., Jintao, D. J., Ying, C., Yixing, Y., & Xingqi, Z. (2010). Impacts of urbanization on hydrology in the Yangtze River Delta, China. *Water Science and Technology, 62,* 1221–1229.

Xu, Y., Zhang, L., & Jia, J. (2008). Lessons from catastrophic dam failures in August 1975 in Zhumadian, China. In *GeoCongress 2008@ sGeosustainability and Geohazard Mitigation, ASCE* (pp. 162–169).

Yue, S., Pilon, P., & Cavadias, G. (2002). Power of the Mann–Kendall and Spearman's rho tests for detecting monotonic trends in hydrological series. *Journal of hydrology, 259*(1), 254–271.

Zhai, P., Zhang, X., Wan, H., & Pan, X. (2005). Trends in total precipitation and frequency of daily precipitation extremes over China. *Journal of Climate, 18,* 1096–1108.

Zhang, Q., Liu, Q., & Wu, L. (2009). Tropical cyclone damages in China 1983–2006. *Bulletin of the American Meteorological Society, 90,* 489–495.

Zhang, S., Guo, Y., & Wang, Z. (2015). Correlation between flood frequency and geomorphologic complexity of rivers network–A case study of Hangzhou China. *Journal of Hydrology, 527,* 113–118.

Zhang, X., Alexander, L., Hegerl, G. C., Jones, P., Tank, A. K., Peterson, T. C., et al. (2011). Indices for monitoring changes in extremes based on daily temperature and precipitation data. *Wiley Interdisciplinary Reviews: Climate Change, 2,* 851–870.

Zhang, Y., Xu, Y., Dong, W., Cao, L., & Sparrow, M. (2006). A future climate scenario of regional changes in extreme climate events over China using the PRECIS climate model. *Geophysical Research Letters, 33,* L24702.

Zheng, Z., Qi, S., & Xu, Y. (2013). Questionable frequent occurrence of urban flood hazards in modern cities of China. *Natural Hazards, 65,* 1009–1010.

Zhou, B., Wen, Q. H., Xu, Y., Song, L., & Zhang, X. (2014). Projected changes in temperature and precipitation extremes in China by the CMIP5 multimodel ensembles. *Journal of Climate, 27,* 6591–6611.

Zhou, H., Wan, J., & Jia, H. (2010). Resilience to natural hazards: A geographic perspective. *Natural Hazards, 53,* 21–41.

Chapter 3
Changes of Water Quality in the Yangtze River Basin

3.1 Introduction

Water scarcity is the lack of freshwater resources to meet the ever-increasing demand for water, which has been the focus of increasing international, national, and local concern and debate (Brunner et al. 2019; Mekonnen and Hoekstra 2016). In recent decades, lots of factors have caused the water scarcity more serious and increased worldwide awareness, which could be summarized into two aspects. One is the increasing water demand, which generally contains the incredible growth in human population (Nielsen 2016); rapid urbanization (Arfanuzzaman and Rahman 2017); competition among agriculture, industry, and cities (or regions) (Duan et al. 2019; Ziolkowska and Peterson 2016); climate change (Gosling and Arnell 2016); and so on. Rapid population growth increases water demand for food production, household consumption, and industrial uses (Liu and Yang 2012; Rasul 2014). Urban growth is also increasing the demand for water resources, and by 2050, urbanization rate is expected to be nearly 70%, which will be likely to consume more water resources (McDonald et al. 2014). With the continuous expansion of a city, people probably find more water from the surrounding region, which will increase competition for freshwater resources with other adjacent cities (Postel 2014). Global climate change has been confirmed to largely affect the hydrological cycle, eventually impacting water resource supplies (Hagemann et al. 2013; Schewe et al. 2014).

The other is the compromised water supply, mainly due to the pollution. That is, water pollution is one of the main causes for the water scarcity (Azizullah et al. 2011). Both anthropogenic activities (such as industrial accidents (Duan et al. 2011; Duan et al. 2013), dam construction (Zhao et al. 2012), and so on) and natural climate change (Delpla et al. 2009), are affecting the water quality everywhere (Chen et al. 2018), resulting in high concentrations of sedimentation, nutrients, heavy metals, and toxic organic compounds, high temperature, and so on (Carr and Neary 2008). For example, as much as 70% of industrial waste and 80% of sewage is directly discharged into the water resources including rivers, lakes, and groundwater

W. Duan and K. Takara, *Impacts of Climate and Human Activities on Water Resources and Quality*, https://doi.org/10.1007/978-981-13-9394-5_3

in developing countries. Thus, it is momentous to detailedly describe spatial–temporal changes in water quality and accurately identify the potential pollution sources (Zhang et al. 2016b).

A series of monitoring programs and protocols have been developed to enable a reliable quantification of nutrient transport in the aquatic system, such as the National Monitoring and Assessment Program (NOVA) in Denmark (Conley et al. 2002; Kronvang et al. 2005), the National Land with Water Information in Japan (Duan et al. 2013), the Harmonized Monitoring Scheme (HMS) in Britain (Hurley et al. 1996; Morvan et al. 2008), the National Water-Quality Assessment (NAWQA) in the United States (Gilliom et al. 1995; Rosen and Lapham 2008), and so on. Meanwhile, lots of approaches including the projection pursuit technique (Zhang and Sihui 2009), neural networks (He et al. 2011), multivariate statistical techniques (Singh et al. 2004) (e.g., discriminant analysis (DA), cluster analysis (CA) and principal component analysis/factor analysis (PCA/FA)), and hydrological models (Moriasi et al. 2012) have been greatly improved to evaluate water quality. Among these methods, multivariate statistical techniques including CA, PCA/FA, and DA can be applied to easily extract important information in large water quality datasets and are therefore used widely to evaluate water quality and identify potential pollution sources (Wang et al. 2013).

Water quality is severely impaired and has a major factor affecting human health and sustainable economic and social development in China (Duan et al. 2018; He et al. 2012). Since 2004, the Ministry of Environmental Protection has begun to pay attention to monitoring systems for surface water quality in the Pearl River Basin, the Tai Lake Basin, the Yellow River Basin, the Yangtze River Basin, the Songhua River Basin, and so on. As a result, a huge monitoring database has been established for regional water resources management, including nutrients, sediments, physical and chemical properties, toxic organic compounds and pesticides, and heavy metals. However, the lake environment is still needed to be evaluated and recovered. For example, Poyang Lake, which is the largest freshwater lake (3050 km^2) in China, is heavily polluted, especially in the year with low water level.

Therefore, this chapter firstly analyzed spatial and temporal distributions of water quality in the Yangtze River Basin, and then focused on the Eastern Poyang Lake Basin to (1) describe temporal and spatial variations of water quality and (2) identify the potential influencing factors that explain changes in water quality parameters using several multivariate statistical approaches (DA, CA, and PCA/FA).

3.2 Materials and Methods

3.2.1 Study Area

Yangtze River begins its journey in the glacial meltwaters of the Tanggula Mountain (elev. 6621 m) and flows 6,397 km eastward before emptying into the Eastern China

Sea (ESC), covering one-fifth of the whole nation with an area of about 1.8×10^6 km^2 (Fig. 3.1b). The river contains an uppermost section (from the source to Yibin, Sichuan Province, with a length of 3,499 km), an upper section (from Yibin to Yichang, Hubei Province, with a length of 1,030 km), a middle section (from Yichang to Hukou, Jiangxi Province, with a length of 950 km), and a lower section (from Hukou to Shanghai, with a length of 938 km). Yangtze River Basin has complicated hydroclimatic conditions due to the East and South Asian monsoon activities. Wet season is from May to October, and dry season is from December to the next April.

Poyang Lake is located on the southern bank of the middle-lower Yangtze River in Jiangxi Province, China (Fig. 3.1c), which is a shallow lake and mainly fed by the Gan River, the Fu River, the Xin River, the Rao River, and the Xiu River. Its drainage area is about 162,200 km^2, 96.7% of which belongs to the Jiangxi Province and 1.8% belongs to Anhui Province. The basin has a subtropical

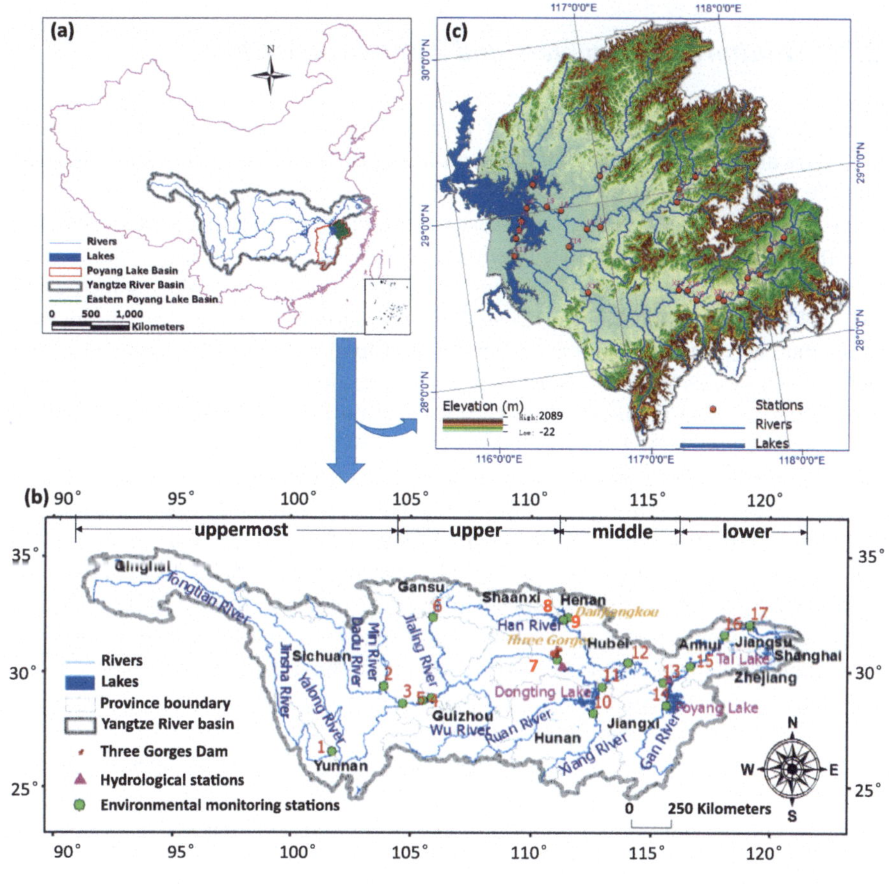

Fig. 3.1 Location of study area and monitoring stations for the rivers in the Yangtze River Basin and the Eastern Poyang Lake Basin

humid climate with short but relatively cool winters and hot and humid summers. In Poyang Lake Basin, the averaged annual regional precipitation and temperature are about 1710 mm and 17.5 °C, respectively.

The Poyang Lake Basin is rich in water resources, but now facing a series of difficulties due to climate change and human activities. With the rapid economic development and population explosions in the basin, human activities including dam construction (Zhang et al. 2012) and land-use change have significantly affected the quality and quantity of water resources (Ye et al. 2013). Therefore, Poyang Lake faces many environmental problems, including deterioration of water quality and eutrophication. In this chapter, the East Poyang Lake Basin (Fig. 3.1) was selected for water quality assessments. The Eastern Poyang Lake Basin mainly contains the Xin River and Rao River (rising from two branches including the Chang River and Lean River).

3.2.2 Monitored Parameters and Analytical Methods

For the Yangtze River Basin, a total of 17 environmental monitoring stations were selected for analysis (Fig. 3.1b). Of them, number 1, 5, 7, 11, 13, 15, and 16 are located along the trunk stream, number 8 and 9 were set up for monitoring the Danjiangkou Reservoir, and the rest were built to assess the influence of major tributaries. Weekly water quality data including pH, chemical oxygen demand (COD_{Mn}), ammonia-nitrogen (NH_4–N), and dissolved oxygen (DO) at these 17 stations during the period 2004–2015 were obtained and processed from a surface water quality monitoring system that was built by China's Ministry of Environmental Protection in 2004. Weekly water grades are also described according to environmental quality standards for surface waters in China (GB3838-2002) (Grade I–V level means that water is "Excellent", "Good", "Satisfactory", "Bad", and "Very bad", respectively).

For the Eastern Poyang Lake Basin, 28 stations were selected for taking water samples; station L1–L9 are located in the Chang River, station X1–X13 are located in the Xin River, station L1–L9 are located in the Chang River, and station P1–P5 are located in East Poyang Lake (see Fig. 3.1). During the period from January 2012 to April 2015, water samples were taken every 2 months. Totally 14 water quality parameters were chosen to describe water quality in the Eastern Poyang Lake Basin, including temperature (TEMP), pH, ammonia–nitrogen (NH_4^+–N), 5-day biochemical oxygen demand (BOD), chemical oxygen demand (COD), dissolved oxygen (DO), total nitrogen (TN), total phosphorus (TP), fluoride (F), sulfide (S), copper (Cu), oil, chromium (Cr), and Zinc (Zn). Table 3.1 shows the specific analytical method for each water quality parameters, which are in line with the standard methods from the State Environment Protection Bureau of China 2002.

Land-use data for the Eastern Poyang Lake Basin from 1980 to 2015 were interpreted from Landsat TM/ETM images with a spatial resolution of 1 × 1 km (Liu et al. 2010) and land-use types contained agricultural land, forest, water body, grassland, construction land, and others.

Table 3.1 Water quality parameters, units, analytical methods, and lowest detected limit as measured from January 2012 to April 2015 for the Eastern Poyang Lake Basin

Parameters	Abbreviations	Units	Analytical methods	Lowest detected limit
Temperature	TEMP	°C	Thermometer	–
pH	pH		Glass electrode	–
Dissolved oxygen	DO	mg/L	Iodimetry	0.2
Ammonia nitrogen	NH_4^+–N	mg/L	*N*-reagent colorimetry	0.05
Biochemical oxygen demand	BOD	mg/L	Dilution and inoculation test	2
Chemical oxygen demand	COD	mg/L	Potassium permanganate method	0.5
Total nitrogen	TN	mg/L	Ultraviolet spectrophotometry	0.05
Total phosphorus	TP	mg/L	Ammonium molybdate spectrophotometry	0.01
Cuprum	Cu	mg/L	Atomic absorption spectrometry	0.001
Zinc	Zn	mg/L	Atomic absorption spectrometry	0.05
Fluoride	F	mg/L	Ion chromatography	0.05
Oil	Oil	mg/L	Infrared spectrophotometry	0.01
Sulfide	S	mg/L	Methylene blue spectrophotometric	0.005
Chromium	Cr	mg/L	DPC colorimetric	0.004

3.2.3 Multivariate Statistical Methods

The seasonal Mann–Kendall test (SMK) (Hirsch and Slack 1984) was applied to detect monotonic trends of weekly water quality data at 17 stations in the Yang River Basin. SMK is a non-parametric test and has been widely used to detect potential trend change points in water quality trends (Bouza-Deaño et al. 2008; Chang 2008; Helsel and Frans 2006; Zhang et al. 2016a). Let $X = (X_1, X_2, \ldots, X_n)^T$ be a time series of independent water quality observations, and $X_i = \left(X_{i1}, X_{i2}, \ldots, X_{ij}\right)$. Here, n is the number of years and j is the number of weekly water quality data for each year and so for weekly "seasons", the first week water quality is compared only with the first week's data of every year, the second week water quality data only with the second week's water quality data of every year, and so on. The null hypothesis (H_0) is that there are no monotonic trends in time. The statistic for the gth season is

$$S_g = \sum_{i=1}^{n-1} \sum_{j=i+1}^{n} sgn\left(X_{jg} - X_{ig}\right), g = 1, 2, \ldots, m \tag{3.1}$$

The SMK statistic, $\hat{S}$, for the entire water quality series can be calculated (Hirsch 1982):

$$\hat{S} = \sum_{g=1}^{m} S_g \tag{3.2}$$

Here, the significance level p is selected at 0.05 and 0.10 with corresponding Z statistics of 1.96 and 1.65, respectively. A positive value of Z represents an "upward trend" and a negative value of Z means a "downward trend".

Spatiotemporal analysis of the water quality in the Eastern Poyang Lake Basin was analyzed by using CA, DA, and PCA/FA techniques. CA is the task of grouping a set of objects based on the characteristics they possess (McKenna 2003; Shrestha and Kazama 2007) and Ward's method is a criterion applied in hierarchical cluster analysis. The Ward's method of hierarchical clustering with squared Euclidean distance was applied to explore the grouping of the 28 sampling stations.

DA determines the variables that discriminate between two or more naturally occurring groups/clusters on the basis of the accuracy rate of discriminant functions (DFs). It constructs a discriminant function (DF) for each group (Johnson and Wichern 1992). DFs are calculated using the following equation:

$$f(G_i) = k_i + \sum_{j=1}^{n} w_{ij} P_{ij} \tag{3.3}$$

where i represents the number of groups (G), k_i represents the constant inherent to each group, n represents the number of parameters, and w_{ij} represents the weight coefficient assigned by DF analysis (DFA) to a given parameter (P_{ij}).

DA was employed to calculate the mean of a variable to predict group membership. The standard, forward stepwise and backward stepwise modes of DA were used to calculate DFs in two groups generated from CA to describe spatial variations in river water quality.

PCA is a dimensionality reduction technique that helps to simplify the data and make it easier to visualize by finding a set of principal components (PCs) (Jolliffe 2002; Yidana et al. 2008). PCs are orthogonal variables calculated by multiplying the original correlated variables with a list of coefficients, which can be described as

$$z_{ij} = a_{i1}x_{1j} + a_{i2}x_{2j} + \ldots + a_{im}x_{mj} \tag{3.4}$$

where z represents the component score, a represents the component loading, x represents the measured value of the variable, i represents the component number, j represents the sample number, and m represents the total number of variables.

FA was used to extract a lower dimensional linear structure from a set of data and therefore provide a powerful means for detecting similarities among samples (Reghunath et al. 2002). FA can reduce the contribution of less significant variables obtained from PCA and the new group of variables known as varifactors (VFs) is extracted through rotating the axis defined by PCA. The basic concept of FA is described as

$$z_{ij} = a_{f1}f_{1j} + a_{f2}f_{21} + \ldots + a_{fm}f_{mi} + e_{fi} \tag{3.5}$$

where z represents the measured value of a variable, a represents the factor loading, f represents the factor score, e represents the residual term accounting for errors or other sources of variation, i represents the number of samples, j represents the number of variables, and m represents the total number of factors.

3.3 Results

3.3.1 *Water Quality in Yangtze River Basin*

3.3.1.1 Level of Water Quality

Figure 3.2 shows changes in different grades of river water at 17 stations in the Yangtze River Basin from 2004 to 2015, suggesting that water environment has been

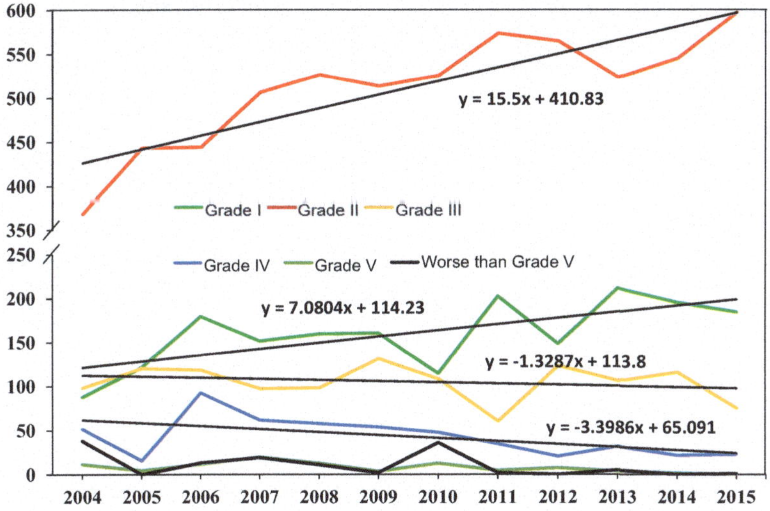

Fig. 3.2 Changes of number of different grades of river water based on 17 stations in the Yangtze River Basin from 2004 to 2015. The black straight lines are the trend lines for each grade

improved significantly since 2004. Concretely, a significant trend was detected in Grade II water, occupying 66.7% of all water quality categories in 2015. A slight increase was found in Grade I water, while a slight decrease in water of Grades III–IV. Proportion of Grade I–III water has increased from 2004 to 2015, with 97.3% in 2015, suggesting gradual improvement in surface water, which is consistent with the results from (Sun et al. 2015).

3.3.1.2 Level of Water Quality

The distribution of pH, DO, COD_{Mn}, and NH_4–N from 2004 to 2015 at 17 stations is shown in the boxplots in Fig. 3.3. As shown in the figure, the median pH and DO concentrations at Stations 1 and 6 were higher compared with other stations, while the median COD_{Mn} and NH_3–N concentrations were lower, which indicates that water quality in the Uppermost Yangtze River Basin was better than in other areas. Except for Station 1, we see an upward trend in median DO and pH from Station 2 to Station 7, but a downward trend in median COD_{Mn} and NH_4–N concentrations. Station 2 (Minjiang Bride) had the highest median COD_{Mn} concentration (3.47 mg/L), the lowest median DO concentration (6.47 mg/L), and relative high NH_4–N concentration. Station 14 (Chuchuo) had the highest median NH_4–N concentration (0.97 mg/L) and the lowest median pH (6.94), followed by station 10 (Xingang), which suggests high ammonia nitrogen pollution in the Xiang River and Gan River.

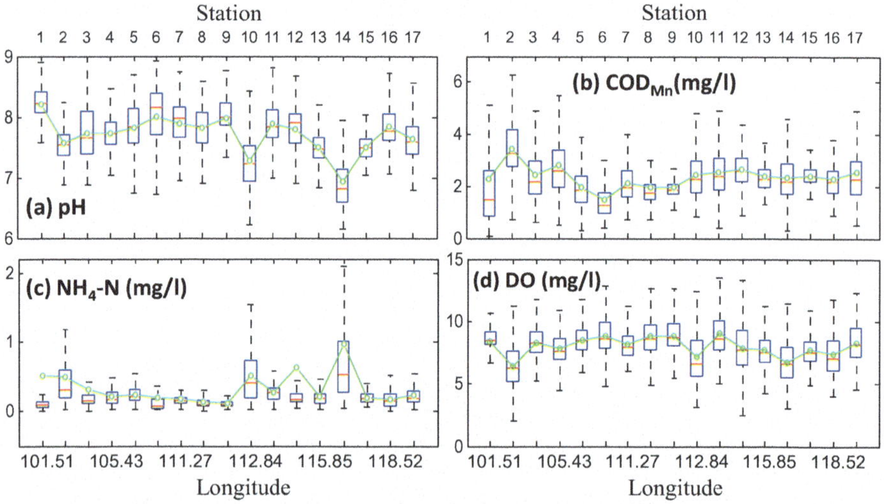

Fig. 3.3 Boxplots illustrating distribution of water quality at 17 stations from 2004 to 2015 in the Yangtze River Basin. Red lines are the mediums. **a** pH, **b** COD_{Mn} (mg/L), **c** NH_4–N (mg/L), **d** DO (mg/L)

3.3.1.3 Trends of Water Quality

Figure 3.4 shows seasonal trends (the value of the test statistic Z) of water quality at 17 stations in the Yangtze River Basin during the period 2004–2015. A decreasing trend was found for pH at 11 stations (approximately 65%, 8 stations with significance at 95% confidence), which were mainly located on the upper and middle of the Yangtze River Basin (Fig. 3.4a). An increasing trend was detected for COD_{Mn} concentration at 10 stations (among them, 9 stations with significance at 95% confidence), which were mainly distributed in major tributaries (e.g., station 10 on Xiang River, station 14 on Gan River) and the lower reach of the trunk stream (Fig. 3.4b). Figure 3.4c indicates that half of the stations had a decreasing trend in NH_4–N concentration, mainly clustering on the middle and lower reaches of the trunk stream. Figure 3.4d shows that a total of 12 stations had an upward trend in DO concentration.

Figure 3.4 also clearly illustrates that there has been a measurable decrease in pollutants before flowing into dams and reservoirs and after flowing from them. For example, water quality after the dams/lakes was better than before the dams/lakes; mean NH_4–N at stations 10 and 14 were 0.53 mg/L and 0.65 mg/L, respectively, but largely decreased to 0.27 mg/L and 0.20 mg/L at stations 11 and 15 after flowing from Dongting and Poyang Lakes. There may be two main reasons for this phenomenon. One is that high runoff from other tributaries with low pollutant concentration directly dilutes outflows from lakes and reservoirs and dams are positively influencing the retention of pollutants along the river. The other is that attenuation of pollutant concentrations because lakes are another well-established explanation for the data (Duan et al. 2015), especially under some appropriate planning and contamination monitoring policies.

To conclude, according to the evaluation of 17 environmental stations, water quality of surface water was significantly improved in the whole Yangtze River Basin during the period 2004–2015. The detailed evaluations were slightly different from

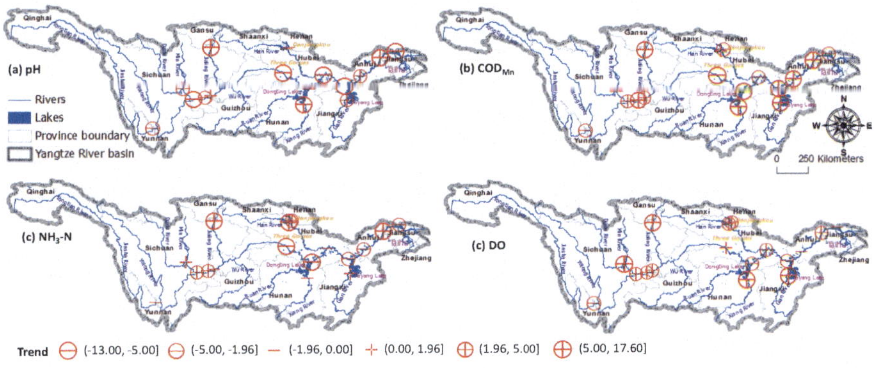

Fig. 3.4 Spatial distribution of seasonal Mann–Kendall trends (the value of Z) **a** pH; **b** COD_{Mn} (mg/L); **c** NH_4–N (mg/L); **d** DO (mg/L) at 17 stations in the Yangtze River Basin, between 2004 and 2015

the State of Environment Report of China (China's Ministry of Environmental Protection 2016), which indicated that Grade I–III water increased slightly from 88.1% in 2014 to 89.4% in 2015, and Grade V+ water remained flat at 3.1%. However, the overall tendency was the same for both the present study and the State of Environment Report. Meanwhile, it is necessary to study and show changes of water quality in some specific regions, such as the Dongting Lake Basin and Poyang Lake Basin.

3.3.2 *Water Quality in Eastern Poyang Lake Basin*

3.3.2.1 Temporal/Spatial Similarity and Grouping

The result of temporal cluster analysis is shown in Fig. 3.5a, which suggests that two groups were detected. Cluster 1 (dry season) contained January and March, representing the low flow period; Cluster 2 (wet season) contained May, July, September, and November, representing the mean and high flow periods. Notably, temporal variation of surface water quality was significantly affected by local climate seasons (spring, summer, autumn, and winter) and hydrological conditions (low, mean, and high flow period). The Poyang Lake Basin lies in a subtropical wet climate zone with a distinct alternation from wet to a dry season, consistent with the temporal patterns of water quality.

Figure 3.5b shows the spatial CA, which also yielded a dendrogram with two statistically significant clusters at $(D_{link}/D_{max}) \times 100 < 60$. The X1 station and the L1–L4

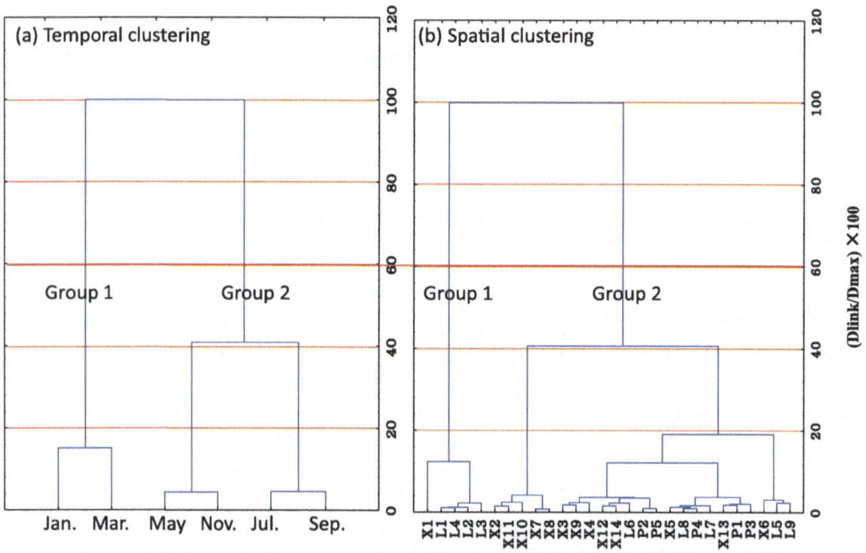

Fig. 3.5 Dendrogram showing the **a** temporal clustering and **b** spatial clustering of study periods

stations in Group 1 are located upstream of Xin River and Le'an River, respectively. Due to low population density and lack of industrial and commercial activities, they are far away from major point and non-point pollution sources. However, the L1–L4 stations are located in Dexing District, which is one of the largest copper and gold production areas in China. Metal pollution and related mineral pollution have always been a big problem (Chen et al. 2015; Xiao et al. 2015). Although relatively high concentrations of Cu, S, and F were observed at the L4 station, the pollution of Group 1 should be considered as the moderate or low level. Group 2 represented the highly polluted sites with the highest average concentration of NH_4–N, petroleum, BOD, COD, and TP. Most of these stations are located in the middle and lower reaches of the Poyang Lake Basin and are contaminated by sources of municipal wastewater, industrial wastewater, and non-point source pollution.

3.3.2.2 Temporal/Spatial Variations in River Water Quality

Tables 3.2 and 3.3 indicate the discriminant functions (DFs) and classification matrices (CMs), which were calculated by the standard, forward stepwise and backward stepwise modes of DA. In the forward stepwise mode, the variables are gradually

Table 3.2 Classification function coefficients for DA of temporal changes

Parameters	Standard mode		Forward stepwise mode		Backward stepwise mode	
	Wet season	Dry season	Wet season	Dry season	Wet season	Dry season
TEMP	1.989	3.033	3.241	2.207	2.613	1.577
pH	82.847	84.585	59.995	58.095	56.048	54.185
NH_4–N	21.105	24.167	28.602	25.452	27.846	25.129
BOD	9.934	10.125				
COD	−1.194	−1.189				
DO	10.981	10.727	12.409	12.687		
TN	−5.506	−7.395	−9.417	−7.549	−12.533	−10.631
TP	7.996	6.712	−3.095	−1.832		
F	46.734	46.243				
S	631.996	634.440				
Cu	−255.025	−250.116				
Oil	−148.788	−206.928	20.532	75.009		
Cr	374.404	403.601				
Zn	53.327	52.592				
Constant	−378.134	−403.979	−302.864	−276.639	−235.610	−206.896

Table 3.3 Classification matrix for DA of temporal changes

Monitoring periods	Percent correct	Temporal groups	
		Wet season	Dry season
Standard mode			
Wet season	95.536	321	15
Dry season	100	0	224
Total	97.321	321	239
Forward stepwise mode			
Wet season	95.536	321	15
Dry season	100	0	224
Total	97.321	321	239
Backward stepwise mode			
Wet season	96.429	324	12
Dry season	100	0	224
Total	97.857	324	236

advanced from the high position until there is no significant change, but in the backward stepwise mode, the gradual deletion is gradually performed from the low position. The standard and forward stepwise mode DFs using 14 and 7 discriminant variables, respectively, produced the corresponding CMs assigning 96.43% of the cases correctly. However, in the backward stepwise mode, the DA used only four discriminant parameters to derive a CM with approximately 97.86% correctly assigned, showing TEMP, pH, NH_4–N, and TN. Therefore, the temporal DA suggested that TEMP, pH, NH_4–N, and TN were the most important parameters to distinguish the difference between the wet season and the dry season, indicating that these four parameters can be used to explain the expected changes in surface water quality in the Eastern Poyang Lake Basin.

Figure 3.6 shows box and whisker plots of the discernible parameters identified by the DA. Due to the local climate, the average temperature in the wet season (Fig. 3.6a) was significantly higher than in the dry season. The same pH difference was found in Fig. 3.6b. Conversely, due to local hydrological conditions, the average concentrations of NH_4–N and TN during the dry season are higher than during the rainy season. The emissions during the rainy season were much larger than during the dry season, which greatly diluted NH_4–N and TN. In addition, during the rainy season (usually in summer and autumn), there are more aquatic organisms than in the dry season, consuming more NH_4–N.

As with temporal DA, the DFs and CMs of the spatial DA were performed using the standard, forward gradual and backward gradual modes according to the space group (low-pollution station and high-pollution station, see Tables 3.4 and 3.5). The standard and forward stepwise mode DFs using 14 and 11 discriminant variables, respectively, produced the corresponding CMs that correctly assigned 95% of the cases, while the backward stepwise DA assigned only 93.75% of the correct assignments to the CMs using only five discriminant parameters (Tables 3.4 and 3.5). The

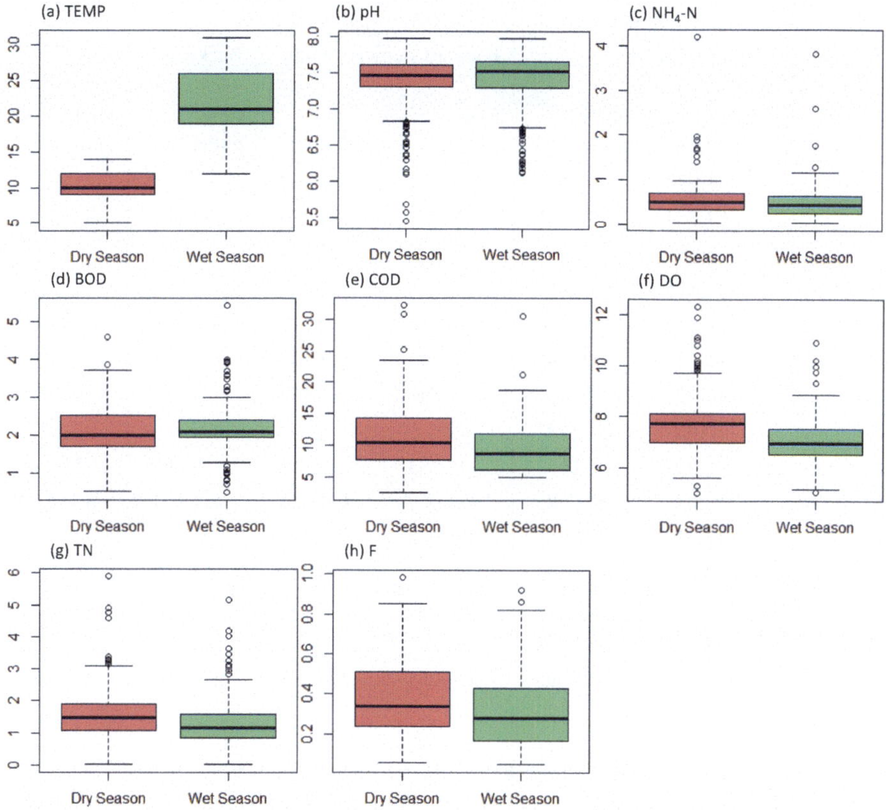

Fig. 3.6 Temporal changes: **a** TEMP, **b** pH, **c** NH_4–N (mg/L), **d** BOD (mg/L), **e** COD (mg/L), **f** DO (mg/L), **g** TN (mg/L), and **h** F (mg/L) in the East Poyang Lake Basin

backward DA indicated that pH, COD, TN, F, and S were the most important parameters to distinguish the difference between a low-pollution station and a high-pollution station.

Figure 3.7 shows the chosen discriminate parameters identified by spatial backward stepwise DA. The pH of the low-pollution areas (Fig. 3.7b) was largely lower than that of the highly polluted areas, which was inconsistent with the analysis of the Danjiangkou Reservoir Basin in China (Chen et al. 2015). This may be because the river section of the area had discharged a large amount of acid mine drainage and copper- and zinc-containing wastewater from the nearby Dexing Copper Mine and many smelters and mining/panning activities. The average COD and TN concentrations in low-pollution areas (Fig. 3.7b, c) were also significantly lower than in highly polluted areas. In highly polluted areas, all sites were located near the middle and lower reaches and near the city, so they were close to municipal sewage and industrial wastewater. The average F and S concentrations in low-pollution areas (Fig. 3.7d, e) were also significantly higher than in highly polluted areas. Obviously, these excess

Table 3.4 Classification function coefficients for DA of spatial changes

Parameters	Standard mode		Forward stepwise mode		Backward stepwise mode	
	Low pollution	High pollution	Low pollution	High pollution	Low pollution	High pollution
TEMP	1.028	1.07	1.214	1.260		
pH	87.554	91.925	79.992	84.230	72.662	77.108
NH_4–N	15.389	13.534	16.827	15.151		
BOD	10.175	10.467				
COD	−0.833	−0.584	−0.409	−0.146	−0.376	−0.146
DO	10.645	10.246	10.144	9.730		
TN	−1.196	0.484	−0.140	1.573	3.971	5.296
TP	10.63	11.57				
F	41.529	37.651	42.883	38.954	59.009	54.412
S	518.07	441.985	479.059	401.226	416.961	330.412
Cu	−244.929	−234.766	−205.655	−195.490		
Oil	−63.009	−43.309	77.283	100.383		
Cr	435.253	496.253	588.508	653.149		
Zn	54.778	55.278				
Constant	−384.511	−413.498	−349.255	−376.823	−273.306	−303.391

acidic contaminants were the main driving force for pH reduction. Figure 3.8 clearly illustrates the spatial distribution of (a) TEMP, (b) pH, (c) NH_4–N (mg/L), (d) BOD (mg/L), (e) COD (mg/L), (f) DO (mg/L), (g) TN (mg/L), (h) TP (mg/L), (i) F (mg/L), (j) S (mg/L), (k) Cu (mg/L), (l) Oil (mg/L), (m) Cr (mg/L), (n) Zn (mg/L) at 27 sites in the eastern basin of Poyang Lake.

3.3.2.3 Data Structure Determination and Source Identification

Based on the normalized log-transformed datasets, PCA/FA was used to further identify the potential pollution sources for the low-pollution and high-pollution regions. Before the PCA/FA analysis, the Kaiser–Meyer–Olkin (KMO) and Bartlett's Sphericity tests were carried out on the parameter correlation matrix to examine the validity of PCA/FA. The KMO results for Group 1 and Group 2 were 0.61 and 0.71, respectively, and Bartlett's Sphericity results were 547.92 and 1611.68 ($p < 0.05$), indicating that the PCA/FA analysis can reasonably reduce size. Figure 3.6 and Table 3.6 indicate that six VFs were calculated for low-pollution areas, and four VFs were calculated for highly contaminated areas with eigenvalues greater than one, accounting for approximately 78.86% and 57.78% of the total variation in the corresponding surface water quality datasets (Fig. 3.9).

Table 3.5 Classification matrix for discriminant analysis of spatial changes

Monitoring stations	Percent correct	Spatial groups	
		Low pollution	High pollution
Standard mode			
Low pollution	72.000	72	28
High pollution	100.000	0	460
Total	95.000	72	28
Forward stepwise mode			
Low pollution	72.000	72	28
High pollution	100.000	0	460
Total	95.000	72	488
Backward stepwise mode			
Low pollution	69.000	69	31
High pollution	99.130	4	456
Total	93.750	73	487

In the low-pollution area, VF1 in the six VFs can explain about 25.38% of the total variance, with strong pH and TN positive loads and moderate COD positive loads, as well as strong S negative loads and medium negative loads of F. The high concentration of total nitrogen reflects agricultural runoff and municipal wastewater (Alexander et al. 2000; Moeder et al. 2017; Yang et al. 2010); chemical oxygen demand is an indicator of organic pollution in industrial and domestic wastewater (Wen et al. 2017). pH is considered to be one of the main reaction conditions involving the redox reaction of organic matter, and the concentration of COD can be adjusted by pH (Juahir et al. 2011). Sulfides and fluorides were mainly derived from copper mines in the region (such as the Dexing Copper Mine in Dexing City), which are very complex and difficult to be used for high sulfur and fluoride beneficiation (He et al. 1997). VF1 included nutrient pollution, organic pollution, and mining pollution. VF2 (14.99% of the total variance) had a strong positive with Cu load and a negative with BOD load, representing metal contamination. VF3 explained 12.21% of the total variance, with a high positive load of Cr and TP and a medium positive load of oil. This factor can be interpreted as representing the effects from point sources such as copper ore, industrial wastewater, and domestic wastewater. VF4, which accounted for 10.49% of the total variance, had a strong positive relationship with temperature and a strong negative relationship with dissolved oxygen load. The concentration of DO is mainly controlled by temperature and therefore has a seasonal and daily cycle (Ringler and Hall 1975). Therefore, due to the low temperature, the dissolved

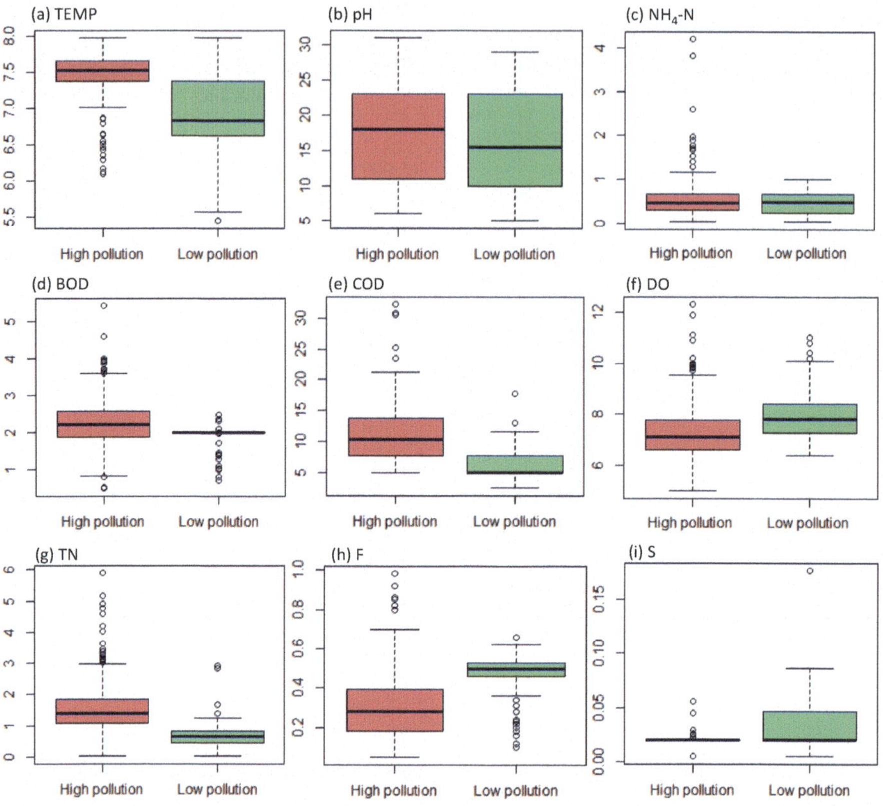

Fig. 3.7 Spatial changes: **a** TEMP, **b** pH, **c** NH_4–N (mg/L), **d** BOD (mg/L), **e** COD (mg/L), **f** DO (mg/L), **g** TN (mg/L), **h** F (mg/L), and **j** S (mg/L) in the East Poyang Lake Basin

oxygen concentration in winter and early spring is high, and the dissolved oxygen concentration in summer and autumn is low due to high temperature. VF5 (9.30%) has a strong positive relationship with NH_4–N load, representing non-point source pollution associated with agricultural activities. VF6 (7.19%) has a strong positive correlation with zinc content, indicating metal contamination.

Regarding the dataset associated with highly polluted areas, VF1, explaining about 22.89% of the total variance in four VFs, had a strong positive load on Cu, and had a moderate negative load on pH and S, which basically represented metal pollution in the upstream. VF2 (16.88% of the total variance) had a strong NH_4–N positive load with moderate TN and COD positive loads. This factor can be interpreted as a typical mixed pollution, including point source pollution (such as industrial and domestic wastewater) and non-point source pollution associated with agricultural activities and atmospheric deposition. VF3 explains 10.19% of the total variance, with a strong DO positive load and a medium positive load of F and a medium negative load of TEMP. Typically, fluoride was from cement plants, fluorination plants, and copper smelters

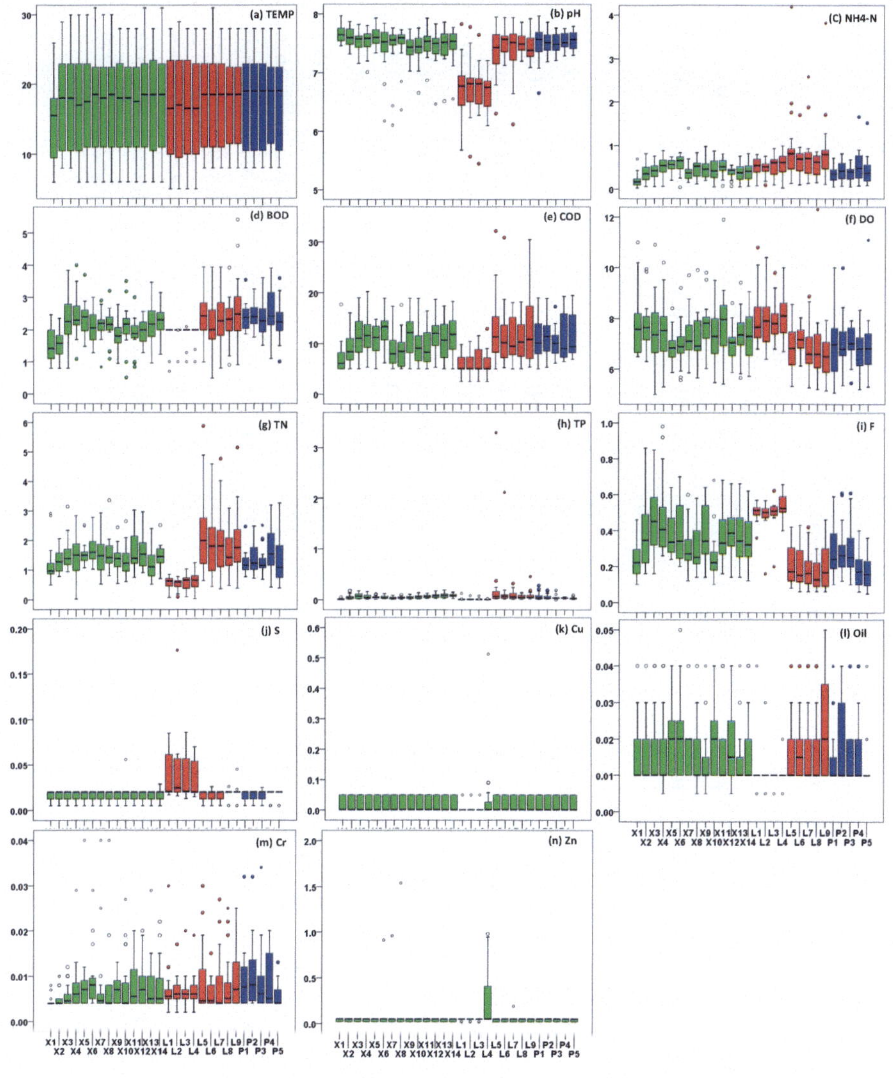

Fig. 3.8 Boxplots illustrating distribution of **a** TEMP, **b** pH, **c** NH_4–N (mg/L), **d** BOD (mg/L), **e** COD (mg/L), **f** DO (mg/L), **g** TN (mg/L), **h** TP (mg/L), **i** F (mg/L), **j** S (mg/L), **k** Cu (mg/L), **l** Oil (mg/L), **m** Cr (mg/L), **n** Zn (mg/L) at 27 stations in the East Poyang Lake Basin

in the region. In low-pollution areas, the relationship between dissolved oxygen and temperature was the same as in the interpretation of VF4. The positive load of BOD of VF4 (7.82%) was very strong, while the load of oil and Cr was medium. High concentrations of BOD and oil may represent organic and petroleum pollution, while cement and copper smelters in the region may produce Cr.

Table 3.6 Loadings of experimental variables (14) on significant VFs for low pollution and high pollution

Parameters	Low pollution (six significant principal components)						High pollution (four significant principal components)			
	VF1	VF2	VF3	VF4	VF5	VF6	VF1	VF2	VF3	VF4
TEMP	0.015	0.026	0.042	0.874	0.211	0.107	0.206	−0.362	−0.623	−0.017
pH	0.730	0.230	0.201	−0.068	−0.243	−0.169	−0.657	−0.149	−0.114	−0.081
NH_4–N	−0.144	−0.078	−0.032	0.084	0.934	0.058	−0.069	0.861	0.022	0.038
BOD	−0.013	−0.834	0.036	0.167	−0.040	0.249	0.103	−0.054	−0.365	0.701
COD	0.607	−0.418	0.362	−0.307	0.263	−0.014	0.050	0.562	0.234	0.318
DO	−0.172	0.197	−0.214	−0.745	0.138	0.165	−0.165	−0.107	0.794	−0.012
TN	0.721	0.283	−0.127	0.186	0.443	−0.124	0.219	0.651	−0.003	0.091
TP	0.149	0.355	0.758	0.097	−0.040	−0.091	0.466	0.413	0.157	0.179
F	−0.597	−0.440	−0.121	−0.253	0.387	0.261	0.330	0.113	0.646	0.065
S	−0.779	0.139	0.067	−0.152	0.155	−0.052	−0.664	0.364	−0.095	−0.178
Cu	0.094	0.741	0.156	−0.023	−0.109	0.334	−0.831	−0.079	0.224	−0.055
Oil	0.420	0.021	0.611	0.165	−0.253	−0.126	0.131	0.202	0.026	0.626
Cr	−0.188	−0.154	0.887	0.075	0.065	0.117	0.073	0.129	0.266	0.639
Zn	−0.123	−0.011	−0.026	−0.017	0.061	0.931	0.835	0.098	−0.169	0.073
Eigenvalue	3.553	2.000	1.709	1.469	1.302	1.007	3.205	2.364	1.426	1.094
Total variance %	25.382	14.287	12.208	10.491	9.303	7.190	22.890	16.883	10.189	7.817
Cumulative variance %	25.382	39.668	51.876	62.367	71.671	78.861	22.890	39.773	49.962	57.779

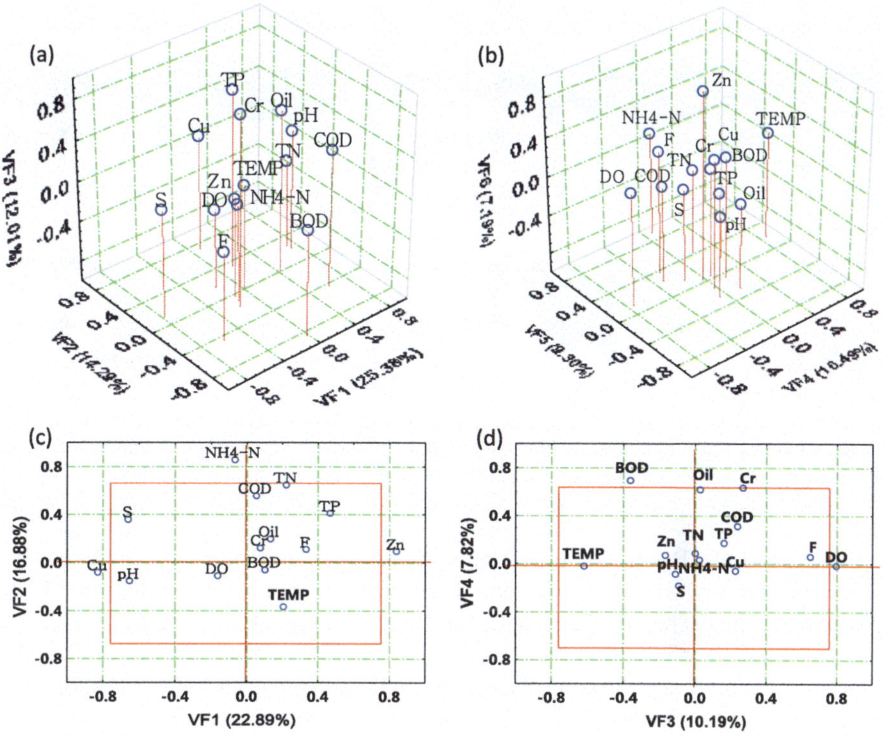

Fig. 3.9 Scatterplot of loadings for the four VFs for group 1 (**a** and **b**) and group 2 (**c** and **d**)

3.4 Discussions

China has been facing severe environmental problems due to rapid population growth and economic development. In recent decades, gross domestic product (GDP) is growing rapidly with the population growth, which brings great downward pressure to water resources. Figure 3.10 shows the spatial distribution of mean concentration at 17 station in Yangtze River Basin, with GDP of 2010, population of 2010, mean precipitation between 2001 and 2010, gross domestic product (GDP), and land use of 2010. As shown in the figure, GDP and population was concentrated in big cities such as the Chengdu–Chongqing Economic Zone along the Upper Yangtze River, the Wuhan–Changsha–Nanchang Economic Zone along the Middle Yangtze River, and the Yangtze River Delta Economic Zone along the Lower Yangtze River, where the water quality was relatively poor at stations 2–4, 10, 14, and 17. The main reason for this phenomenon was the increase of pollutants from municipal sewage, agricultural wastewater, and livestock production facilities in these big cities (Wang et al. 2008; Zhang et al. 2010). Also, enormous effort has been made in recent decades in China including an ever-improving legal system and the popularity of sewage treatment plants, which significantly improve water quality in Yangtze River Basin.

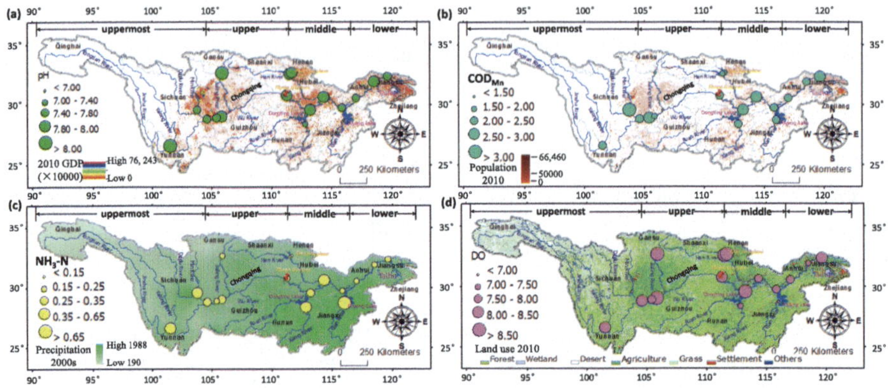

Fig. 3.10 Spatial distributions of mean concentrations **a** pH; **b** COD_{Mn} (mg/L); **c** NH_3–N (mg/L); and **d** DO (mg/L) at 17 stations, with GDP of 2010 (Chinese yuan/km^2), population of 2010 (person/km^2), mean precipitation between 2001 and 2010 (mm), and land use of 2010

In Poyang Lake Basin, five potential pollutants including nutrients, organic matter, chemicals, heavy metals, and natural pollutants were identified. Regional high pollution may be the result of drainage of municipal wastewater, agricultural wastewater, and livestock production facilities (Xue et al. 2019). Water quality is strongly influenced by land use through changes in pollutant input and water circulation. On the one hand, with the growth of agriculture and settlements, like other countries, point source pollution and non-point source pollution may increase due to the increase in sewage treatment plants and the additional use of agricultural fertilizers. On the other hand, land use can accelerate the transport of pollutants by changing the form of runoff. For example, most of the rainwater that falls on the parking lot will flow out immediately (Duan et al. 2016), often draining into the rainwater pipeline, and the pollutants in the sewage will be transported to the stream or ditches without filtration. Figure 3.11 shows changes of land use in the east Poyang Lake Basin from 1980 to 2015, which suggests that an increasing trend was found in agricultural land and construction land, probably deteriorating water quality in this area.

3.5 Conclusions

This chapter firstly characterized the spatiotemporal distribution, long-term trend, and seasonality of water quality in the Yangtze River Basin based on weekly water quality data (pH, COD, NH_4-N, and DO) from 17 environmental stations for the period January 2004 through December 2015, and then focused on the Eastern Poyang Lake Basin to systematically evaluate the spatiotemporal variation and identify the potential influencing factors for surface water quality. Results indicate that water quality was gradually improved during this time period in the Yangtze River Basin, but regional differences were still obvious. In the Poyang Lake Basin, the

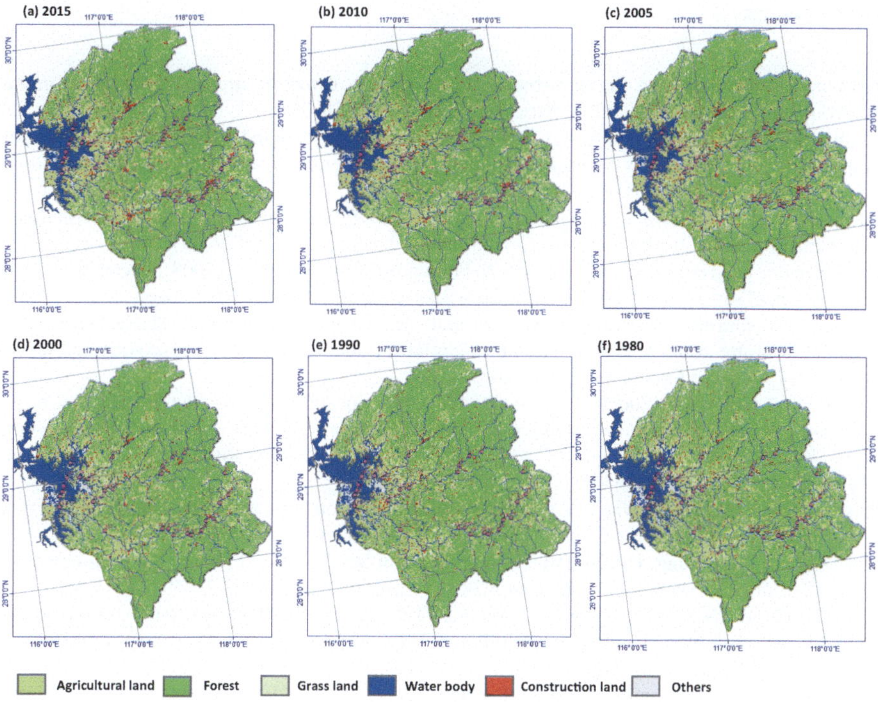

Fig. 3.11 Land-use changes in the East Poyang Lake Basin from 1980 to 2015

hierarchical CA divided 6 months into two groups based on similar water quality characteristics, and also divided 28 sampling stations into two groups, which can provide a basis for reasonable and useful classification, thereby optimizing the space monitoring network at a lower cost in the near future. For temporal changes, four discriminant parameters including TEMP, pH, NH_4–N, and TN were identified, with about 97.86% correct assignations. Five discriminant parameters including pH, COD, TN, F, and S were identified, with about 93.75% correct assignations. All these methods can be used to optimize water quality monitoring procedures in future. The results of the PCA/FA analysis identified five potential contaminations, including nutrients, organic matter, chemicals, heavy metals, and natural pollution. Copper mines such as Dexing Copper Mine and land-use change probably caused high pollution in the Eastern Poyang Lake Basin.

References

Alexander, R. B., Smith, R. A., & Schwarz, G. E. (2000). Effect of stream channel size on the delivery of nitrogen to the Gulf of Mexico. *Nature, 403,* 758–761.

Arfanuzzaman, M., & Rahman, A. A. (2017). Sustainable water demand management in the face of rapid urbanization and ground water depletion for social-ecological resilience building. *Global Ecology and Conservation, 10,* 9–22.

Azizullah, A., Khattak, M. N. K., Richter, P., & Häder, D. (2011). Water pollution in Pakistan and its impact on public health-a review. *Environment International, 37,* 479–497.

Bouza-Deaño, R., Ternero-Rodríguez, M., & Fernández-Espinosa, A. J. (2008). Trend study and assessment of surface water quality in the Ebro River (Spain). *Journal of Hydrology, 361,* 227–239.

Brunner, M. I., Gurung, A. B., Zappa, M., Zekollari, H., Farinotti, D., & Stähli, M. (2019). Present and future water scarcity in Switzerland: Potential for alleviation through reservoirs and lakes. *Science of the Total Environment, 666,* 1033–1047.

Carr, G. M., & Neary, J. P. (2008). Water quality for ecosystem and human health (2nd ed.). In: *United Nations Environment Program Global Environment Monitoring System (GEMS)/Water Program.*

Chang, H. (2008). Spatial analysis of water quality trends in the Han River Basin, South Korea. *Water Research, 42,* 3285–3304.

Chen, H., Teng, Y., Lu, S., Wang, Y., & Wang, J. (2015a). Contamination features and health risk of soil heavy metals in China. *Science of the Total Environment, 512,* 143–153.

Chen, P., Li, L., & Zhang, H. (2015b). Spatio-Temporal Variations and Source Apportionment of Water Pollution in Danjiangkou Reservoir Basin, Central China. *Water, 7,* 2591–2611.

Chen, W., He, B., Nover, D., Duan, W., Luo, C., Zhao, K., et al. (2018). Spatiotemporal patterns and source attribution of nitrogen pollution in a typical headwater agricultural watershed in Southeastern China. *Environmental Science and Pollution Research, 25,* 2756–2773.

China's Ministry of Environmental Protection 2016. (2015). State of Environment Report. Retrieved May 25, 2017, from http://www.zhb.gov.cn/gkml/hbb/qt/201606/W020160602413860519309.pdf.

Conley, D. J., Markager, S., Andersen, J., Ellermann, T., & Svendsen, L. M. (2002). Coastal eutrophication and the Danish national aquatic monitoring and assessment program. *Estuaries and Coasts, 25,* 848–861.

Delpla, I., Jung, A., Baures, E., Clement, M., & Thomas, O. (2009). Impacts of climate change on surface water quality in relation to drinking water production. *Environment International, 35,* 1225–1233.

Duan, W., Chen, G., Ye, Q., & Chen, Q. (2011). The situation of hazardous chemical accidents in China between 2000 and 2006. *Journal of Hazardous Materials, 186,* 1489–1494.

Duan, W., Chen, Y., Zou, S., & Nover, D. (2019). Managing the water-climate-food nexus for sustainable development in Turkmenistan. *Journal of Cleaner Production, 220,* 212–224.

Duan, W., He, B., Chen, Y., Zou, S., Wang, Y., Nover, D., et al. (2018). Identification of long-term trends and seasonality in high-frequency water quality data from the Yangtze River Basin, China. *PLoS ONE, 13,* e188889.

Duan, W., He, B., Nover, D., Fan, J., Yang, G., Chen, W., et al. (2016). Floods and associated socioeconomic damages in China over the last century. *Natural Hazards, 82,* 401–413.

Duan, W., He, B., Takara, K., Luo, P., Nover, D., Sahu, N., et al. (2013a). Spatiotemporal evaluation of water quality incidents in Japan between 1996 and 2007. *Chemosphere, 93,* 946–953.

Duan, W., Takara, K., He, B., Luo, P., Nover, D., & Yamashiki, Y. (2013b). Spatial and temporal trends in estimates of nutrient and suspended sediment loads in the Ishikari River, Japan, 1985 to 2010. *Science of the Total Environment, 461,* 499–508.

Duan, W. L., He, B., Takara, K., Luo, P. P., Nover, D., & Hu, M. C. (2015). Modeling suspended sediment sources and transport in the Ishikari River Basin, Japan, using SPARROW. *Hydrology and Earth System Sciences, 19,* 1293–1306.

Gilliom, R. J., Alley, W. M., & Gurtz, M. E. (1995). *Design of national water-quality assessment program.* United States Government Printing Office.

Gosling, S. N., & Arnell, N. W. (2016). A global assessment of the impact of climate change on water scarcity. *Climatic Change, 134,* 371–385.

Hagemann, S., Chen, C., Clark, D. B., Folwell, S., Gosling, S. N., Haddeland, I., et al. (2013). Climate change impact on available water resources obtained using multiple global climate and hydrology models. *Earth System Dynamics, 4,* 129–144.

He, B., Oki, T., Sun, F., Komori, D., Kanae, S., Wang, Y., et al. (2011). Estimating monthly total nitrogen concentration in streams by using artificial neural network. *Journal of Environmental Management, 92,* 172–177.

He, G., Lu, Y., Mol, A. P., & Beckers, T. (2012). Changes and challenges: China's environmental management in transition. *Environmental Development, 3,* 25–38.

He, M., Wang, Z., & Tang, H. (1997). Spatial and temporal patterns of acidity and heavy metals in predicting the potential for ecological impact on the Le An river polluted by acid mine drainage. *Science of the Total Environment, 206,* 67–77.

Helsel, D. R., & Frans, L. M. (2006). Regional Kendall test for trend. *Environmental Science and Technology, 40,* 4066–4073.

Hirsch, R. M. (1982). A comparison of four streamflow record extension techniques. *Water Resources Research, 18*(4), 1081–1088.

Hirsch, R. M., & Slack, J. R. (1984). A nonparametric trend test for seasonal data with serial dependence. *Water Resources Research, 20,* 727–732.

Hurley, M. A., Currie, J. E., Gough, J., & Butterwick, C. (1996). A framework for the analysis of harmonised monitoring scheme data for England and Wales. *Environmetrics, 7,* 379–390.

Johnson, R. A., & Wichern, D. W. (1992). *Applied multivariate statistical analysis*. NJ: Prentice hall Englewood Cliffs.

Jolliffe, I. (2002). *Principal component analysis*. New York: Springer-Verlag.

Juahir, H., Zain, S. M., Yusoff, M. K., Hanidza, T. T., Armi, A. M., Toriman, M. E., et al. (2011). Spatial water quality assessment of Langat River Basin (Malaysia) using environmetric techniques. *Environmental Monitoring and Assessment, 173,* 625–641.

Kronvang, B., Jeppesen, E., Conley, D. J., Søndergaard, M., Larsen, S. E., Ovesen, N. B., et al. (2005). Nutrient pressures and ecological responses to nutrient loading reductions in Danish streams, lakes and coastal waters. *Journal of Hydrology, 304,* 274–288.

Liu, J., & Yang, W. (2012). Water sustainability for China and beyond. *Science, 337,* 649–650.

Liu, J., Zhang, Z., Xu, X., Kuang, W., Zhou, W., Zhang, S., et al. (2010). Spatial patterns and driving forces of land use change in China during the early 21st century. *Journal of Geographical Sciences, 20,* 483–494.

McDonald, R. I., Weber, K., Padowski, J., Flörke, M., Schneider, C., Green, P. A., et al. (2014). Water on an urban planet: Urbanization and the reach of urban water infrastructure. *Global Environmental Change, 27,* 96–105.

McKenna, J. E. (2003). An enhanced cluster analysis program with bootstrap significance testing for ecological community analysis. *Environmental Modelling and Software, 18,* 205–220.

Mekonnen, M. M., & Hoekstra, A. Y. (2016). Four billion people facing severe water scarcity. *Science Advances, 2,* e1500323.

Moeder, M., Carranza-Diaz, O., López-Angulo, G., Vega-Aviña, R., Chávez-Durán, F. A., Jomaa, S., et al. (2017). Potential of vegetated ditches to manage organic pollutants derived from agricultural runoff and domestic sewage: A case study in Sinaloa (Mexico). *Science of the Total Environment, 598,* 1106–1115.

Moriasi, D. N., Wilson, B. N., Douglas-Mankin, K. R., Arnold, J. G., & Gowda, P. H. (2012). Hydrologic and water quality models: use, calibration, and validation. *Transactions of the Asabe, 55,* 1241–1247.

Morvan, X., Saby, N., Arrouays, D., Le Bas, C., Jones, R., Verheijen, F., et al. (2008). Soil monitoring in Europe: a review of existing systems and requirements for harmonisation. *Science of the Total Environment, 391,* 1–12.

Nielsen, R. W. (2016). Growth of the world population in the past 12,000 years and its link to the economic growth. *Journal of Economics Bibliography, 3,* 1–12.

Postel, S. (2014). The last oasis: facing water scarcity. Routledge.

Rasul, G. (2014). Food, water, and energy security in South Asia: A nexus perspective from the Hindu Kush himalayan region☆. *Environmental Science & Policy, 39,* 35–48.

Reghunath, R., Murthy, T. S., & Raghavan, B. R. (2002). The utility of multivariate statistical techniques in hydrogeochemical studies: an example from Karnataka, India. *Water Research, 36,* 2437–2442.

Ringler, N. H., & Hall, J. D. (1975). Effects of logging on water temperature, and dissolved oxygen in spawning Beds. *Transactions of the American Fisheries Society, 104,* 111–121.

Rosen, M. R., & Lapham, W. W. (2008). Introduction to the US geological survey national water-quality assessment (NAWQA) of ground-water quality trends and comparison to other national programs. *Journal of Environmental Quality, 37,* 190.

Schewe, J., Heinke, J., Gerten, D., Haddeland, I., Arnell, N. W., Clark, D. B., et al. (2014). Multi-model assessment of water scarcity under climate change. *Proceedings of the National Academy of Sciences, 111,* 3245–3250.

Shrestha, S., & Kazama, F. (2007). Assessment of surface water quality using multivariate statistical techniques: A case study of the Fuji river basin, Japan. *Environmental Modelling and Software, 22,* 464–475.

Singh, K. P., Malik, A., Mohan, D., & Sinha, S. (2004). Multivariate statistical techniques for the evaluation of spatial and temporal variations in water quality of Gomti River (India)—a case study. *Water Research, 38,* 3980–3992.

Sun, J., Chen, Y., Zhang, Z., Wang, P., Song, X., Wei, X., et al. (2015). The spatio-temporal variations of surface water quality in China during the "Eleventh Five-Year Plan". *Environmental Monitoring and Assessment, 187,* 1–12.

Wang, J., Da, L., Song, K., & Li, B. (2008). Temporal variations of surface water quality in urban, suburban and rural areas during rapid urbanization in Shanghai, China. *Environmental Pollution, 152,* 387–393.

Wang, Y., Wang, P., Bai, Y., Tian, Z., Li, J., Shao, X., et al. (2013). Assessment of surface water quality via multivariate statistical techniques: A case study of the Songhua River Harbin region, China. *Journal of Hydro-Environment Research, 7,* 30–40.

Wen, Y., Schoups, G., & Van De Giesen, N. (2017). Organic pollution of rivers: Combined threats of urbanization, livestock farming and global climate change. *Scientific Reports, 7,* 43289.

Xiao, X., Wang, Y., Zhang, H., & Yu, X. (2015). Effects of primary productivity and ecosystem size on food-chain length in Raohe River, China. *Acta Ecologica Sinica, 35,* 29–34.

Xue, H., Sayama, T., Takara, K., He, B., Huang, G., & Duan, W. (2019). Non-point source pollution estimation in the Pingqiao River Basin, China, using a spatial hydrograph-separation approach. *Hydrological Sciences Journal, 64,* 962–973.

Yang, Y., Zhou, F., Guo, H., Sheng, H., Liu, H., Dao, X., et al. (2010). Analysis of spatial and temporal water pollution patterns in Lake Dianchi using multivariate statistical methods. *Environmental Monitoring and Assessment, 170,* 407–416.

Ye, X., Zhang, Q., Liu, J., Li, X., & Xu, C. (2013). Distinguishing the relative impacts of climate change and human activities on variation of streamflow in the Poyang Lake catchment, China. *Journal of Hydrology, 494,* 83–95.

Yidana, S. M., Ophori, D., & Banoeng-Yakubo, B. (2008). A multivariate statistical analysis of surface water chemistry data-The Ankobra Basin, Ghana. *Journal of Environmental Management, 86,* 80–87.

Zhang, C., & Sihui, D. (2009). A new water quality assessment model based on projection pursuit technique. *Journal of Environmental Sciences, 21,* S154–S157.

Zhang, Q., Li, L., Wang, Y. G., Werner, A. D., Xin, P., Jiang, T., et al. (2012a). Has the Three-Gorges Dam made the Poyang Lake wetlands wetter and drier? *Geophysical Research Letters, 39,* L20402.

Zhang, Y., Cabilio, P., & Nadeem, K. (2016a). Improved seasonal Mann–kendall tests for trend analysis in water resources time series. In *Advances in time series methods and applications* (pp. 215–229). Springer.

Zhang, Z., Juying, L., Mamat, Z., & QingFu, Y. (2016b). Sources identification and pollution evaluation of heavy metals in the surface sediments of Bortala River, Northwest China. *Ecotoxicology and Environmental Safety, 126,* 94–101.

Zhang, Z., Tao, F., Du, J., Shi, P., Yu, D., Meng, Y., et al. (2010). Surface water quality and its control in a river with intensive human impacts-a case study of the Xiangjiang River, China. *Journal of Environmental Management, 91,* 2483–2490.

Zhao, Q., Liu, S., Deng, L., Yang, Z., Dong, S., Wang, C., et al. (2012). Spatio-temporal variation of heavy metals in fresh water after dam construction: a case study of the Manwan Reservoir, Lancang River. *Environmental Monitoring and Assessment, 184,* 4253–4266.

Ziolkowska, J. R., & Peterson, J. M. (2016). *Competition for water resources: experiences and management approaches in the US and Europe* (p. 478). Cambridge: Elsevier.

Chapter 4
Spatiotemporal Evaluation of Water Quality and Water Quality Incidents Over Japan

4.1 Introduction

Water is one of the leading drivers of sustainable development of human social economy (Steffen et al. 2015). With the fast development of economy and the growth of population, the demand of freshwater for human is growing rapidly, and water shortage is one of the major crises of the twenty-first century in the world (Pangare 2006; Pearce 2018; Polizzotto et al. 2008). The world population is more than 7.3 billion people in 2015 and is expected to be up to 8.5 billion by 2030, 9.7 billion in 2050, and 11.2 billion in 2100, which will consume more resources (e.g., food, freshwater, and so on) and produce vast quantities of waste, causing pollution of many lakes and rivers (Vörösmarty et al. 2010). Meanwhile, increasing population and economy bring more pollutants, which are climate change such as global warming and extreme precipitation events have significantly affected water resources and water quality, exacerbating the water problem (Vörösmarty et al. 2000). Therefore, how to improve water quality and reduce the risk of pollution incidents has become a very important and urgent issue for controlling water pollution.

In Japan, due to strict emission regulations, development of sewage treatment systems, and other related measures in recent years, water quality management has made great efforts to improve river water quality (Jun et al. 2004; Luo et al. 2011; Takahasi 2009). However, freshwater supply still faces serious waste problems (Honma and Hu 2009; Nakano et al. 2008) and some studies have been done for improving water quality (Hosono et al. 2010; Tsujimoto et al. 2006; Yasuhara and Yamazaki 2005). Some researches argued that the occurrence of water pollution incidents is an important cause of the deterioration of surface water quality (Wakakura and Iiduka 1999), and few studies have investigated the characteristics of water quality events (such as spatial and temporal distribution).

Therefore, this chapter firstly describes the trend and distribution of water quality and water quality incidents in Japan, from aspects of incident numbers, pollutant

W. Duan and K. Takara, *Impacts of Climate and Human Activities on Water Resources and Quality*, https://doi.org/10.1007/978-981-13-9394-5_4

category, incident cause, and affected extend. Then, reasons for the rapid growth have also been discussed as the types of pollutants change and the cause of the incident.

4.2 Materials and Methods

4.2.1 Regional Divisions

Japan is an archipelago consisting of four large islands (Honshu, Hokkaido, Kyushu, and Shikoku) and many other small islands (of which 6,852 islands have a coastline of more than 100 m). Japan is also made up of many short and fast rivers, which are divided into three levels including the first-class river systems, the second-class river systems, and other river systems. Generally, the first-class river systems have 109 rivers with corresponding lengths of 87,560 km, and the second-class river systems have 2723 rivers with corresponding lengths of 36010 km. The first-class river systems are charged by the Ministry of Land, Infrastructure and Transportation of Japan (MLIT), the second-class river systems are controlled by the government of urban and rural prefectures, and other small rivers are managed by Mayors of municipalities. Figure 4.1 shows nine regions for river management in Japan, including Hokkaido, Tohoku, Kanto, Hokuriku, Chubu, Kinki, Chugoku, Shikoku, and Kyushu, which suggests that Kanto has the longest channel length, up to 13181.8 km, followed by Tohoku and Kinki.

4.2.2 Datasets

Yearly water quality data in rivers, lakes and reservoirs, and sea and coastal areas from 1974 to 2017 were obtained from the Ministry of the Environment (MOE) of Japan. Every year, the MOE put together the results of the nationwide water quality survey of public water areas implemented by the national and local governments. Location of the main 98 water quality monitoring stations is shown in Fig. 4.2. Of them, there are 12 stations in Hokkaido, 11 stations in Tohoku, 8 stations in Kanto, 10 stations in Hokuriku, 11 stations in Chubu, 10 stations in Kinki, 10 stations in Chugoku, 5 stations in Shikoku, and 21 stations in Kyushu. In this chapter, the water quality indicator mainly contains biochemical oxygen demand (BOD), chemical oxygen demand (COD), total nitrogen (TN), and total phosphorus (TP).

Water pollution incident data was obtained from the Japan River Water Quality Yearbook (1996–2017), which was edited and issued by the Japan Rivers Association (JRA) and the MLIT. With these statistical data, we could find and determine the distribution of pollutant categories, the causes and losses involved in the accident, and the trend of pollution accidents in Japan. However, in this chapter, the detailed

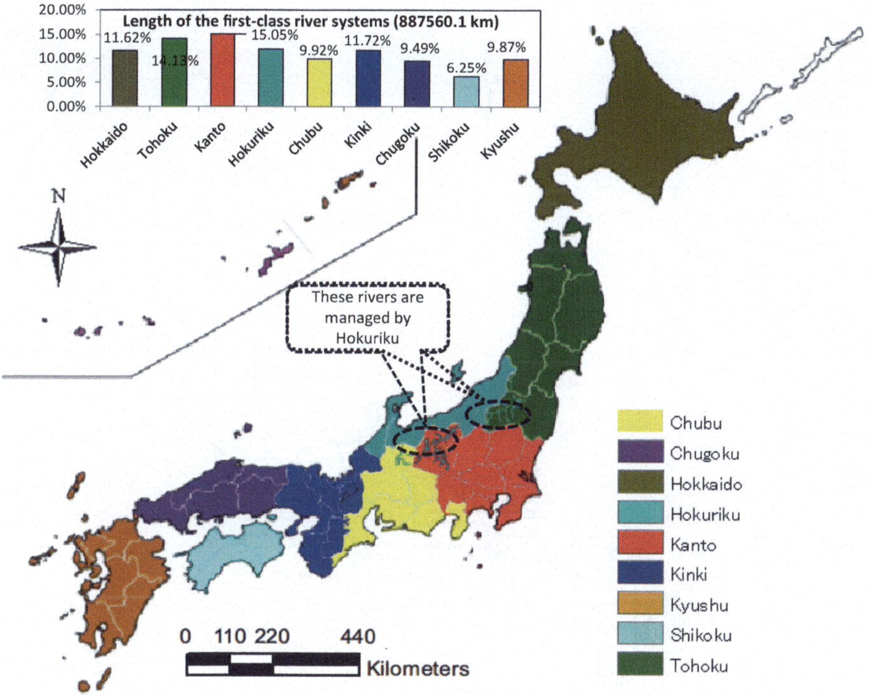

Fig. 4.1 Nine regions for river management in Japan (Modified from MLIT)

pollutant categories for the incidents were from 1996 to 2007. Length of the rivers for the three types of river systems is checked and obtained from the MLIT (http://www.mlit.go.jp/en/index.html) and JRA (http://www.japanriver.or.jp). Water supply facilities were obtained from the Health Services Bureau of the Ministry of Health, Labor and Welfare (MHLW) (http://www.mhlw.go.jp/english). Illegal dumping was found from the Ministry of the Environment of Japan (http://www.env.go.jp/en/).

4.2.3 Methods

Trends for water quality indicators and water pollution incidents were estimated by linear least-squares regression, and its significance in each time series was estimated by the Mann–Kendall trend test (Kendall 1975). The linear regression could be expressed by

$$Y_i = \beta X_i + \varepsilon_i \tag{4.1}$$

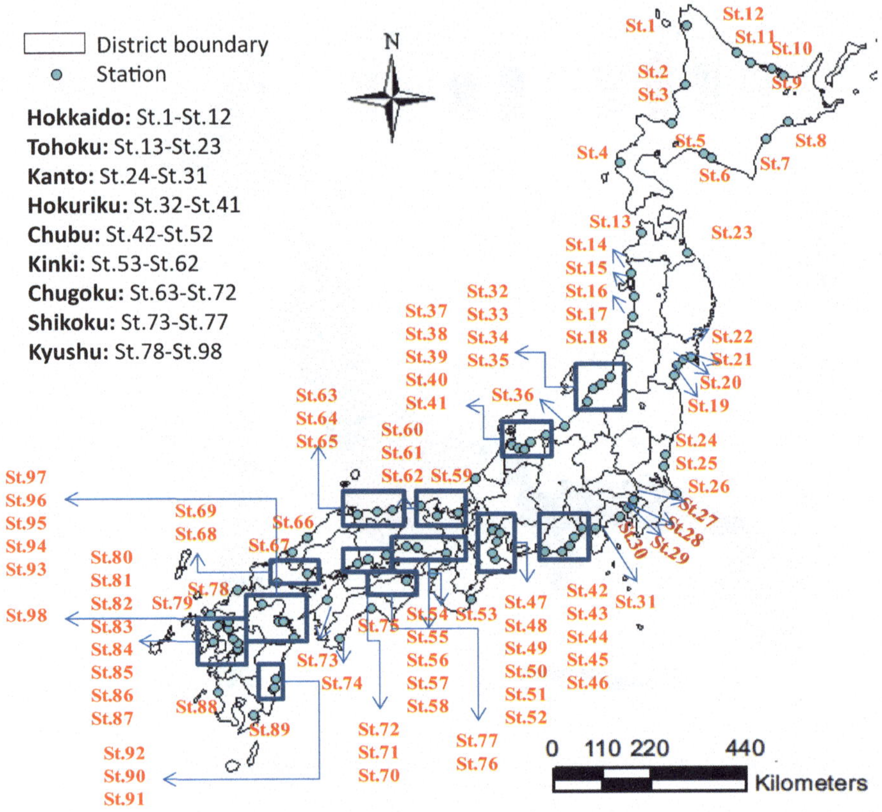

Fig. 4.2 Location of 98 water quality monitoring stations in nine regions in Japan. The order of station ID is demonstrated from up to down or from right to left in each square (Modified from Luo et al. 2011)

where Y_i represents a dependent variable (response) for ith experimental/sampling unit, X_i represents the level of the independent (predictor) variable for ith experimental/sampling unit, β represents the change in mean of Y when X increases by 1 (slope), and ε_i is the random error term.

The water pollution incidents were analyzed from the frequency of events and the number of pollutant types, the number of related causes, long-term trends, and geographical distribution. Here, statistics were used to show the spatiotemporal characteristics of incidents that were the result of time lapses or regional differences. The reasons for these features were also discussed.

4.3 Results and Discussion

4.3.1 Changes of Water Quality

Figure 4.3 shows changes in compliance rate of environmental quality standards (hereinafter referred to as "the EQSs", the details can be accessed from the MOE) for living environment items in Japan from 1974 to 2017. Both BOD and COD are typical water quality indicators for organic contamination. In terms of the compliance rates of EQSs for BOD and COD (BOD for rivers and COD for lakes and reservoirs and sea and coastal areas) in 2017, the overall compliance rate was 89%, with rivers at 94%, lakes and reservoirs at 53.2%, and sea areas at 78.6%. A significant increasing linear trend was detected in BOD for rivers and COD for lakes and reservoirs, while a slight decreasing trend was found in COD for sea and coastal areas. The year of 2015 had the highest compliance rates in total, rivers, land lakes, and reservoirs, with the corresponding values at 90.3, 95.8, and 58.7%.

Figure 4.4 shows changes of BOD and COD concentrations in Japan from 1979 to 2017, suggesting that a significant increasing trend was found for BOD concentration in rivers and COD concentration in lakes and reservoirs, while a slight decreasing trend for COD concentration in sea and coastal areas. These trends were the same as the compliance rate of environmental quality standards. The BOD concentration in rivers decreased from 3.3 mg/L in 1979 to 1.2 mg/L in 2017, more than a threefold fall. The COD concentration in lakes and reservoirs decreased from 4.2 mg/L in 1979 to 3.2 mg/L in 2017, almost a quarter fall. The COD concentration in sea and coastal areas fluctuated between 1.7 mg/L and 1.9 mg/L during the period 1979–2017. Figures 4.2 and 4.3 indicate an especially remarkable improvement in the compliance rate of EQSs in Japan during recent decades.

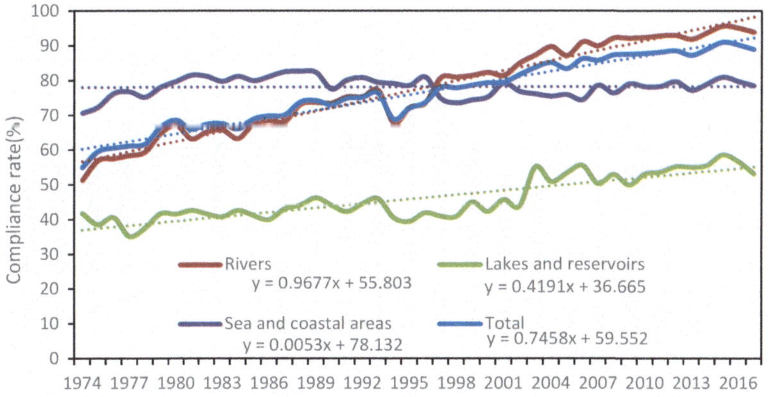

Fig. 4.3 Changes in compliance rate of environmental quality standards for BOD and COD (BOD for rivers and COD for lakes and reservoirs and sea and coastal areas) for living environment items in Japan from 1974 to 2017. The dashed lines are linear trend lines

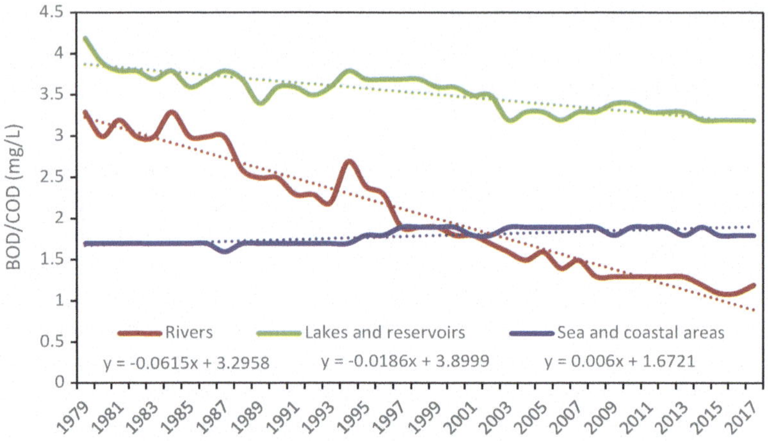

Fig. 4.4 Changes in BOD in rivers, and COD in lakes and reservoirs, and sea and coastal areas in Japan from 1979 to 2017. The dashed lines are linear trend lines

4.3.2 Changes of the Incidents Over Time

Figure 4.5 shows the nationwide trend of Japan and the total number of water quality incidents fluctuated between 400 and 1600/year, and the average from 1996 to 2017, from 1996 to 1999, from 2000 to 2003, from 2004 to 2007, and from 2008 to 2017 were 1080, 499, 902, 1487, and 1222, respectively. Obviously, the average from 2004 to 2007 was almost three times as many as from 2000 to 2003, which means water pollution incidents were almost three times increasing from 1996 to 2007. But the number decreased from 1487 (2004–2007) to 1222 (2008–2017). The largest number

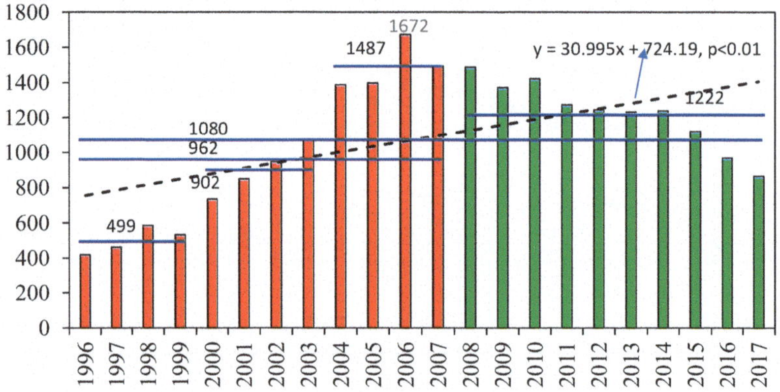

Fig. 4.5 Number of water pollution incidents in Japan from 1996 to 2017

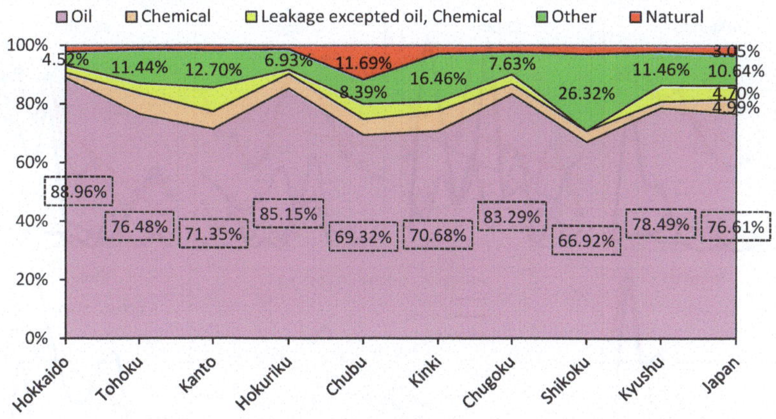

Fig. 4.6 Distribution of water pollution incidents by category from 1996 to 2007

of water quality incidents was found in 2006, up to 1672, followed by 2007 (1493) and 2008 (1488). Generally, water quality incidents firstly increased from 1996 to 2007, and then decreased from 2008 to 2017.

4.3.3 Results of Pollutant Category

4.3.3.1 Distribution by Pollutant Category

In order to determine the type of water pollution incidents, pollutant categories were recorded using the fire descriptor including "Oil", "Chemical", "Spills excepted oil and chemical", "Other", and "Natural". The "Other" category includes incidents for which the cause has not been determined. The "Nature" category includes incidents that are not in the artificial category. Figure 4.6 indicates the national distribution and regional distribution of pollution incidents by pollutant category for total incidents from 1996 to 2007, which suggests that oil was the largest proportion of pollutant types in Japan, occupying 76.61%. The oil category was also the main pollution source for each region, and Hokkaido had the largest oil category pollution incidents, up to 88.96%, followed by Hokuriku (85.15%) and Chugoku (83.29%). Shikoku had the lowest proportion (0%) for the "natural" category, while Chubu had the largest proportion, up to 11.69%.

4.3.3.2 Trend in Pollutant Category

Figure 4.7 shows the trend of water pollution incidents attributed to each category for the nine regions from 1996 to 2007, suggesting that there has been much change

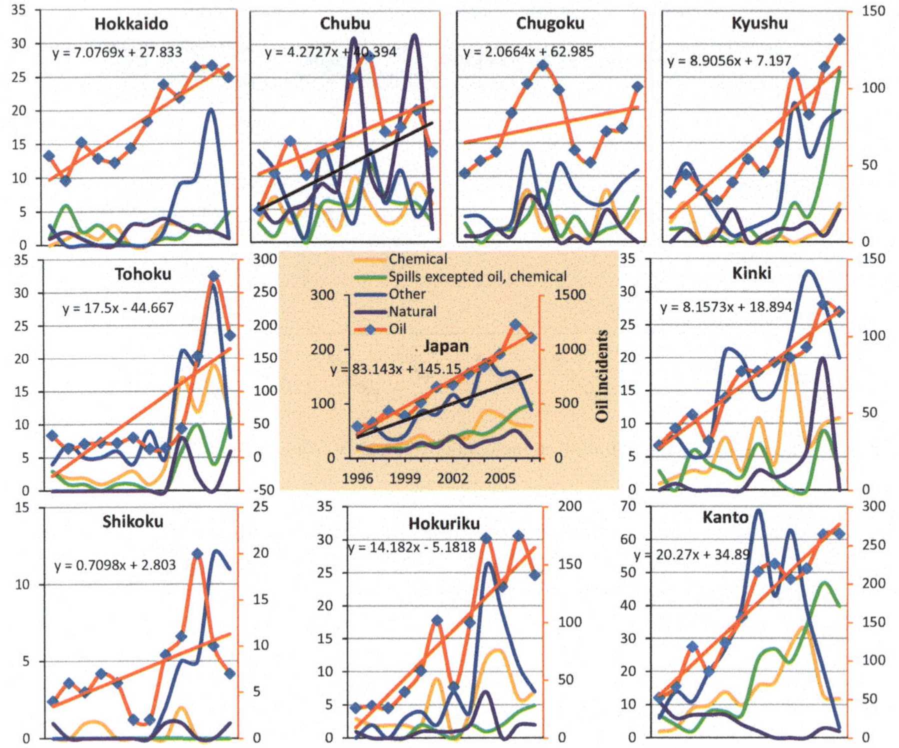

Fig. 4.7 Trends in pollutant category. The red and black straight lines show the trend of pollution incidents caused by "oil" category and "Natural" category, respectively

in the category of pollution incidents. A growing trend was found for the number of each category, especially since 2005, and "Oil" and "Other" categories fluctuated violently during the study period. Kanto had the largest trend for the "Oil" category, up to 20 incidents/year. Chubu had the largest growth for the "Natural" category, which was in line with Fig. 4.6.

4.3.4 Results of Cause

4.3.4.1 Cause of Incidents

In order to identify the cause of a water pollution incident, seven descriptors including "Poor working practice", "Equipment failure", "Accident", "Illegal dumping", "Other", "Unknown", and "Natural" were used to document the cause of the contaminant. The cause of "accident" refers to traffic accident. The cause of "Other" refers to accidents other than factories and traffic accidents, such as aviation accidents. The

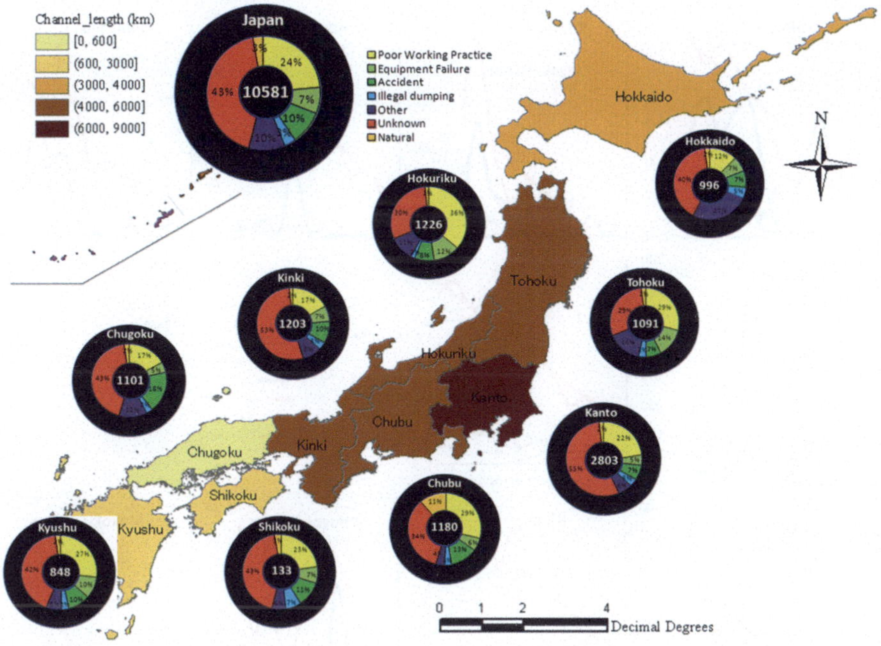

Fig. 4.8 Distribution of water pollution incidents by cause ("Poor working practice", "Equipment failure", "Accident", "Illegal dumping", "Other", "Unknown", and "Natural") during the study period

cause of "Unknown" includes events that cannot determine the cause, such as unexplained oil spills, chemical leaks, etc. The cause of "Natural" includes events that are not man-made. According to statistics from 10,581 incidents in Japan, Fig. 4.8 indicates that the main cause of the incident was "unknown" (43%), followed by "Poor work practices" (24%), "Accident" (10%), and "Other" (10%). Figure 5.7 also shows that in Hokuriku, the main cause of the incident was "Poor work practice" (36%), followed by "unknown" (30%) and "Equipment failure" (12%); in Tohoku, "Poor work practice" and "Unknown" had the same proportion (29%), and Chubu had the largest proportion for "Accident" cause, up to 11%, which was in line with Fig. 4.7.

4.3.4.2 Trend in Incident Cause

Figure 4.9 shows the time series and trends of the incident causes from 1996 to 2007 in Japan. Generally, the number of incidents from each cause had an increasing trend during the period 1996–2017, especially from the "Unknown" cause and "Poor

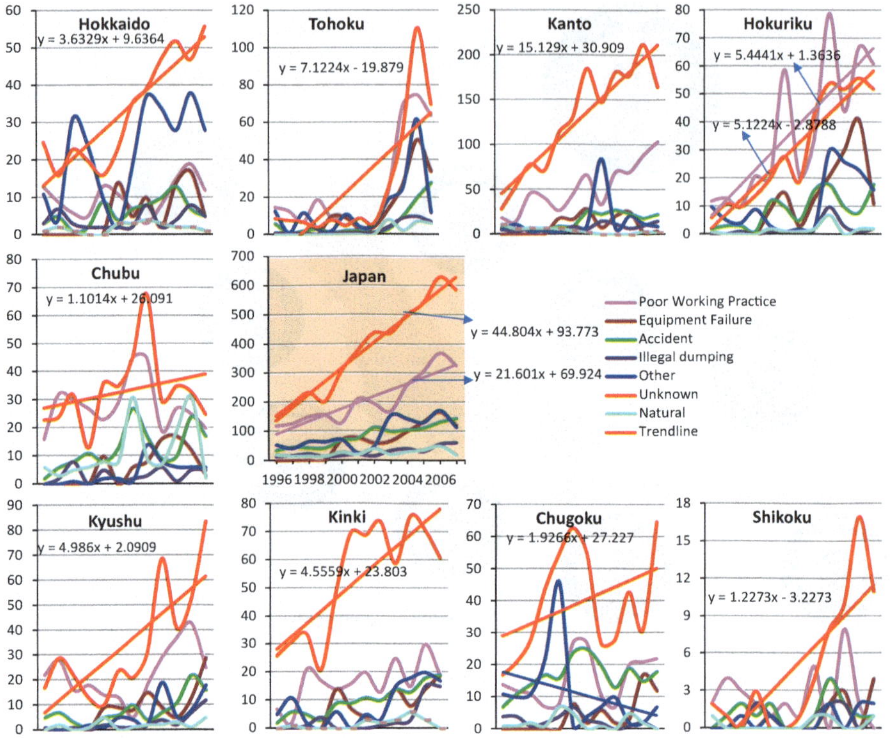

Fig. 4.9 Causes of pollution incidents. The red, pink, and blue straight lines show the trend of pollution incidents caused by "Unknown" cause, "Poor working practice" cause, and "Other" cause, respectively

working practice" cause (Fig. 4.9 (Japan)), which largely reflects the changes for all nine sub-regions (Fig. 4.9 (Hokkaido) and (Shikoku)). Meanwhile, some special cases were found for the trend in incident cause. For example, Fig. 4.9 (Chugoku) shows a decreasing tendency of incidents from the "Other" cause in Chugoku.

4.4 Impacts of Water Pollution

It is obvious that water pollution incidents have significant negative effects on people, local communities, and wildlife, and many years are needed to recover (Exley and Esiri 2006; Nsiah-Kumi 2008).

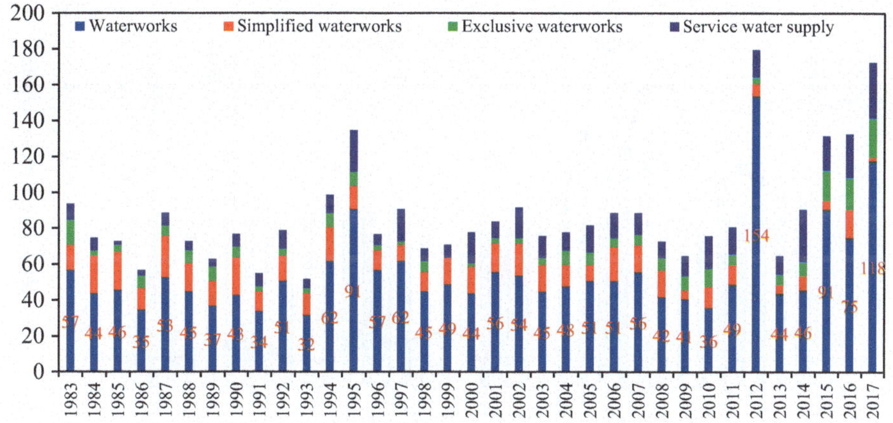

Fig. 4.10 Changes in four kinds of water supply establishments under the impact of water pollution incidents from 1983 to 2017

4.4.1 Changes in Different Water Supply Establishments

In Japan, water supply is divided into four categories (2012). Service water supply: a utility that supplies water to a water company through a water pipe. Waterworks: a population of 5001 people or more according to the water supply plan. Simplified waterworks: a population of 101 or more but less than 5,000 people based on the water supply plan. Exclusive waterworks: a water plant owned by a population of 101 or more people according to the water supply plan. Figure 4.10 shows the changes in the four water supply establishments under the influence of water pollution incidents from 1983 to 2017. As can be seen from the figure, the most affected one was the waterworks in these years, about 60% per year, followed by the simplified waterworks. In 2012, the number of water pollution incidents induced by the waterworks was up to 154, which was the largest value during the period 1983–2017. The main reason for this phenomenon is that the annual water supply of the waterworks (for example, according to the 2002 statistics of 15.745 billion cubic meters, accounting for 94.8%) was largely greater than other types of water supply establishments, which probably increased the risk of water pollution.

4.4.2 Water Supply by Offensive Tastes and Odors

The changes in the incidence of water affected by offensive tastes and odors in the water supply from 1983 to 2017 are shown in Fig. 4.11. As can be seen in the figure, the number of utilities by offensive odors and tastes peaked maintained a relatively higher and constant level, and the year 2016 had the largest value, up to 135. The population suffering from offensive tastes firstly increased from 1983 to 1990, and

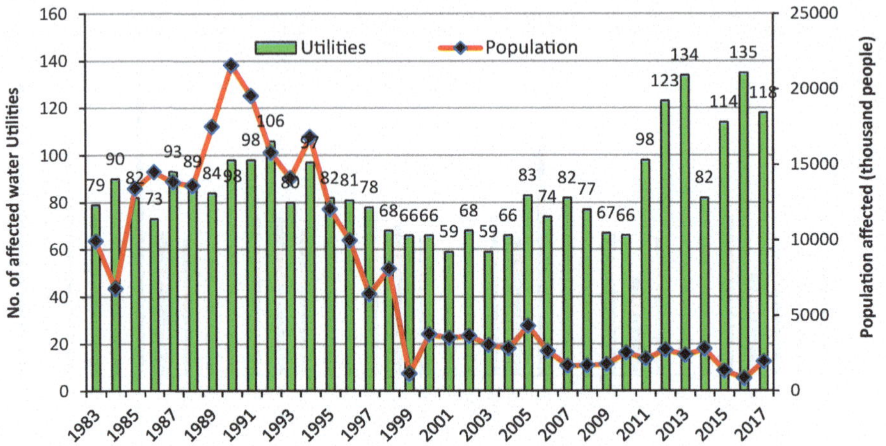

Fig. 4.11 Changes in the incidence of offensive tastes and odors in water supply system from 1983 to 2017

then decreased since 1990. The population was up to 21.63 million in 1990, which was the highest value during the period 1983–2017. The population suffering from offensive tastes in 2017 was about 1.98 million, an increase from about 860,000 in 2016. The number of water utilities that suffered offensive tastes was 118, a decrease from 135 in 2016. The investment of water treatment technologies such as ozonation was the reason of decrease of people since 1990 (Ando 2008).

4.5 Discussions

Among the results of pollutant category before, oil spill is no doubt the biggest risk resulting in water quality incidents and it is also the main problem in other countries. The reasons for this phenomenon are mainly as follows: Firstly, Japan is the third largest oil consumer after the United States and China. Although the proportion of oil consumption decreased somewhat with the spread of energy saving and fuel conversion, it was still about half of energy supply. In addition, oil spill, release of a liquid petroleum hydrocarbon into the environment because of human activity, is an inevitable problem at the same time. The type of oil spill, perhaps surprising to many, is that oil from individual cars and boats, lawn mowers, underground pipeline, refinery, and airplanes contributes the most oil pollution to the river in Japan (Yudakuken 2012). Except oil, chemical and other substances spill also contributed to growing incidence of water pollution.

The above results indicate that oil spills were undoubtedly the biggest risk of water quality incidents and also a major problem in other countries (Shadizadeh and Zoveidavianpoor 2010; Wang et al. 2003). The main reasons for this phenomenon are as follows: First, Japan is the third largest oil consumer after the United States and

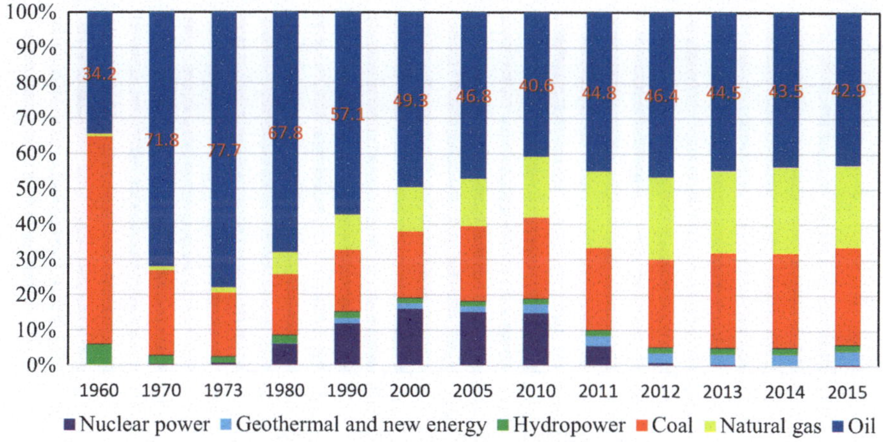

Fig. 4.12 The primary energy supply trends in Japan (%), and the source data was from the Agency for National Resources and Energy of Japan

China. Although the proportion of oil consumption has declined with the spread of energy conservation and fuel conversion, it still accounts for about half of the energy supply (see Fig. 4.12). In addition, due to the wrong human activities, the leakage of liquid petroleum hydrocarbons into the environment is an inevitable problem. The type of oil spill may surprise many people by the fact that oil from individual cars and boats, lawn mowers, underground pipelines, refineries, and aircraft has the greatest oil pollution to Japanese rivers (Yudakuken 2012). Besides oil, changes in leakage of chemicals and other substances have also become two factors that contribute to the growth of water quality accidents.

In addition, from the perspective of the cause of the incident, the causes of leakage of oil, chemicals, and other substances are mainly "Unknown", "Poor working practice", "Equipment failure", and "Illegal dumping". Among them, "Poor working practice" can lead to erroneous operations, which directly lead to incidents (Bentley and Haslam 2001) and have appeared in many industries (Diane et al. 1999; Mearns et al. 2003). In recent decades, it has become a serious problem in Japan. The number of non-regular workers (such as dispatched workers, fixed-term contract workers, and part-time workers, who generally lack systematic safety education and operation) is increasing among employed people (as shown in Fig. 4.13) due to a decline in birth rates.

Illegal dumping is another major problem that has caused water quality incidents in Japan (Ichinose and Yamamoto 2011; Miyazaki and Une 2005). Although the number of cases has declined since 2001, the weight of illegal dumping has not decreased correspondingly (as shown in Fig. 4.14). Most of them have a huge negative impact on water resources management. For example, some give off awful smells from toxic gases created from chemical reactions of the waste materials, some often gases are quite poisonous, even deadly, all of which pose a risk to improve water quality. Finally, natural disasters also threaten water quality. For example, the Great East

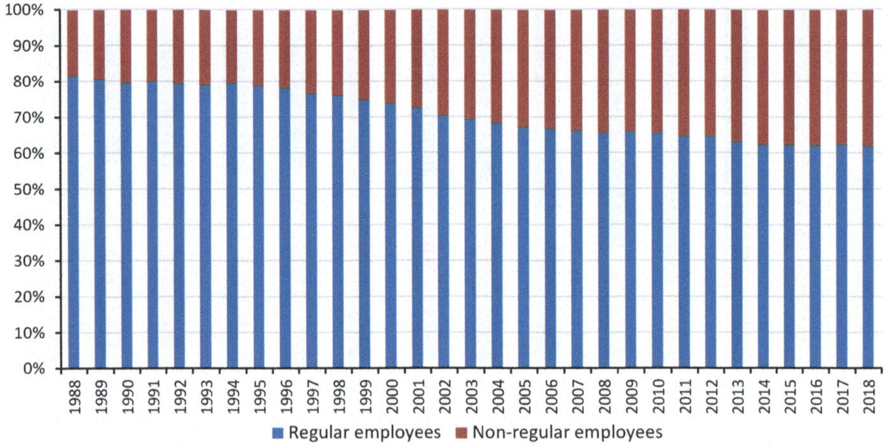

Fig. 4.13 Proportion of non-regular and regular employees from 1988 to 2018 in Japan (*Data source* Labour force survey 2019)

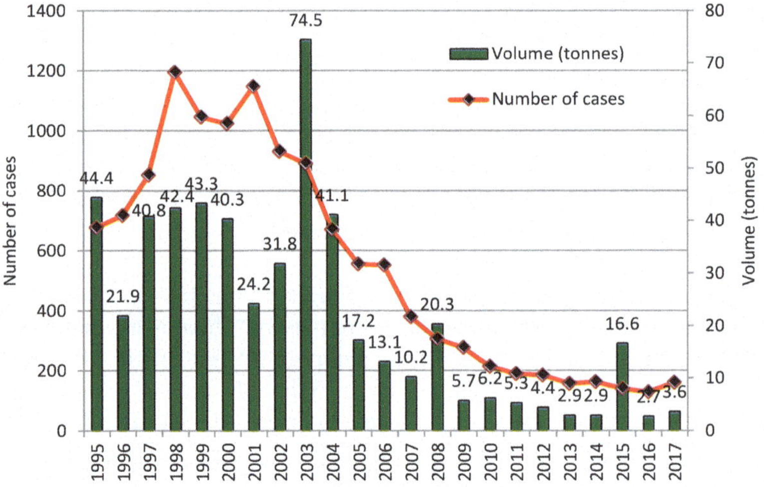

Fig. 4.14 Trend in illegal dumping from 1995 to 2017 (*Data source* Ministry of the environment of Japan 2019)

Japan Earthquake in the Tohoku region hit the sewer system, and many wastewater treatment plants were unable to process it, which had a negative impact on water quality. In addition, the nuclear leakage after the Great Tohoku Earthquake caused serious water pollution problems, which have not yet been resolved.

4.6 Conclusions

Some conclusions obtained are as follows. Firstly, significant improvement was found in water quality in recent decades in Japan, and the overall compliance rate was up to 89% in 2017. Secondly, the number of water quality incidents increased from 1996 to 2007, and then decreased since 2008. According to the analysis of pollutant categories, in addition, oil was the largest proportion of Japan's pollutant categories (76.61%) and was the main pollution source of each region. Of the 10,581 incidents that occurred in Japan between 1996 and 2007, the main cause was "Unknown" (43%), followed by "Poor work practice" (24%), followed by "Accident" (10%), and "Others" (10%). Waterworks (approximately 60%) had the largest value in the four water supply establishments under the influence of water pollution incidents from 1983 to 2017. Finally, the number of non-regular workers has been increasing and it becomes a big problem causing the water quality incidents.

References

Ando, S. (2008). History and development of drinking water quality management in Japan. Retrieved May 2019, from http://www.jwrc-net.or.jp/aswin/projects-activities/rd_files/jp-kr_cooperation/2008_soul100th_es.pdf.

Bentley, T. A., & Haslam, R. A. (2001). A comparison of safety practices used by managers of high and low accident rate postal delivery offices. *Safety Science, 37,* 19–37.

Diane, E., A. G. F. GIBB & Roger, A. H. (1999). The quality of accident and health data in the construction industry: interviews with senior managers. *Construction Management and Economics, 17*(2), 197–204.

Exley, C., & Esiri, M. M. (2006). Severe cerebral congophilic angiopathy coincident with increased brain aluminium in a resident of Camelford, Cornwall, UK. *Journal of Neurology, Neurosurgery and Psychiatry, 77,* 877–879.

Honma, S., & Hu, J. L. (2009). Efficient waste and pollution abatements for regions in Japan. *International Journal of Sustainable Development and World Ecology, 16,* 270–285.

Hosono, T., Su, C. C., Okamura, K., & Taniguchi, M. (2010). Historical record of heavy metal pollution deduced by lead isotope ratios in core sediments from the Osaka Bay, Japan. *Journal of Geochemical Exploration, 107,* 1–8.

Ichinose, D., & Yamamoto, M. (2011). On the relationship between the provision of waste management service and illegal dumping. *Resource and Energy Economics, 33,* 79–93.

Jun, T., Hiroaki, T., Yuji, O., & Yutaka, S. (2004). River Management and Countermeasure Technologies for Accidental Spills by Ministry of Land, Infrastructure and Transport in Japan.

Kendall, M. G. (1975). *Rank correlation methods*, 4th ed.; Charles Griffin: London, UK, 1975; ISBN 978-0-85-264199-6.

Labour Force Survey. (2019). Statistics Bureau of Japan. Retrieved June, 2019, from https://www.stat.go.jp/english/data/roudou/.

Luo, P., He, B., Takara, K., Razafindrabe, B. H. N., Nover, D., & Yamashiki, Y. (2011). Spatiotemporal trend analysis of recent river water quality conditions in Japan. *Journal of Environmental Monitoring, 13,* 2819–2829.

Mearns, K., Whitaker, S. M., & Flin, R. (2003). Safety climate, safety management practice and safety performance in offshore environments. *Safety Science, 41,* 641–680.

Ministry of the Environment of Japan. (2019). Report of illegal dumping (in Japanese). Retrieved June, 2019, from https://www.env.go.jp/recycle/ill_dum/santouki/.

Miyazaki, M., & Une, H. (2005). Infectious waste management in Japan: A revised regulation and a management process in medical institutions. *Waste Management, 25,* 616–621.

Nakano, T., Tayasu, I., Yamada, Y., Hosono, T., Igeta, A., Hyodo, F., et al. (2008). Effect of agriculture on water quality of Lake Biwa tributaries, Japan. *Science of the Total Environment, 389,* 132–148.

Nsiah-Kumi, P. A. (2008). Communicating effectively with vulnerable populations during water contamination events. *Journal of Water and Health, 6,* 63–76.

Pangare, G. (2006). The source of the problem. *Nature, 441,* 28.

Pearce, F. (2018). *When the rivers run dry, fully revised and updated edition: Water-the defining crisis of the twenty-first century*, Beacon Press.

Polizzotto, M. L., Kocar, B. D., Benner, S. G., Sampson, M., & Fendorf, S. (2008). Near-surface wetland sediments as a source of arsenic release to ground water in Asia. *Nature, 454,* 505–508.

Shadizadeh, S. R., & Zoveidavianpoor, M. (2010). Assessment of petroleum leakage in groundwater of the Abadan refinery. *Journal of Water and Wastewater, 74,* 27–36.

Steffen, W., Richardson, K., Rockström, J., Cornell, S. E., Fetzer, I., Bennett, E. M., et al. (2015). Planetary boundaries: Guiding human development on a changing planet. *Science, 347,* 1259855.

Takahasi, Y. (2009). History of water management in japan from the end of world war II. *Water Resources Development, 25,* 547–553.

Tsujimoto, A., Nomura, R., Yasuhara, M., Yamazaki, H., & Yoshikawa, S. (2006). Impact of eutrophication on shallow marine benthic foraminifers over the last 150 years in Osaka Bay, Japan. *Marine Micropaleontology, 60,* 258–268.

Vörösmarty, C. J., Green, P., Salisbury, J., & Lammers, R. B. (2000). Global water resources: Vulnerability from climate change and population growth. *Science, 289,* 284.

Vörösmarty, C. J., McIntyre, P. B., Gessner, M. O., Dudgeon, D., Prusevich, A., Green, P., et al. (2010). Global threats to human water security and river biodiversity. *Nature, 467,* 555.

Wakakura, M., & Iiduka, Y. (1999). Trends in chemical hazards in Japan. *Journal of Loss Prevention in the Process Industries, 12,* 79–84.

Wang, H., Wang, C., Wu, W., Mo, Z., & Wang, Z. (2003). Persistent organic pollutants in water and surface sediments of Taihu Lake, China and risk assessment. *Chemosphere, 50,* 557–562.

Yasuhara, M., & Yamazaki, H. (2005). The impact of 150 years of anthropogenic pollution on the shallow marine ostracode fauna, Osaka Bay, Japan. *Marine Micropaleontology, 55,* 63–74.

Yudakuken. Cases of oil spill (in Japanese). Retrieved May 2019, from http://www.yudakuken.com/jikorei.html.

Chapter 5
Assessment of Precipitation Amounts and Climate Extremes in Japan

5.1 Introduction

Changes in extreme weather and climate extremes have a major negative impact on the natural environment and human society and are one of the most serious challenges to society in addressing climate change (2008). For example, rapid declines in rainfall may have serious impacts on regional ecosystems and human settlements and may lead to social decline (e.g., the Mayan civilization) (Medina-Elizalde and Rohling 2012) and biodiversity reduction (e.g., Amazonia in 2005) (Marengo et al. 2008). It is important to fully understand the precipitation changes in the past for predicting what will happen for the extreme precipitation event in the future, which could be used to improve the ability to manage the risks of water-related disasters (Duan et al. 2019). Many studies have focused on large-scale spatial-scale rainfall changes. Some studies have found an increase in global extreme precipitation (Allan and Soden 2008; Coumou and Rahmstorf 2012; Easterling et al. 2000; Lenderink and Van Meijgaard 2008), while others have detected an increasing trend in droughts due to global warming (Burns et al. 2010; Dai 2011; Kerr 2007).

Due to the impact of climate change, increasing trends in precipitation extreme events have been found in Japan (Duan et al. 2015; Hu et al. 2015; Solomon 2007). For example, based on 50 gauged stations, Fujibe et al. (2005, 2006) found that three extreme precipitation indices including extreme daily precipitation, extreme four-hourly and hourly precipitation increased in the past century over Japan. With changes in the topographic distribution, Miyajima and Fujibe (2011) argued that a moderate north–south gradient appeared for the distribution of top ten-minute precipitation. Some other researchers have analyzed changes in precipitation at different time scales and regions across Japan (Iwasaki and Sunaga 2009; Takeshita 2010).

All of these studies have detected the characteristics of precipitation, but there has been little information on precipitation changes and extreme precipitation events in the last century. Therefore, this chapter calculates the temporal and spatial variations

W. Duan and K. Takara, *Impacts of Climate and Human Activities on Water Resources and Quality*, https://doi.org/10.1007/978-981-13-9394-5_5

of seasonal changes in precipitation amounts and extreme events over Japan in the past century and explores the effect of sea surface temperature on extreme precipitation events.

5.2 Data and Methods

5.2.1 Datasets and Quality Control

Figure 5.1 and Table 5.1 show the location and information of 51 observed weather stations in Japan, which were developed by the Japan Meteorological Agency (JMA) and used to gather sunlight, temperature, precipitation, and wind velocity and direction (Kawamoto et al. 2011). The daily precipitation from 1901 to 2012 was used to compute interannual and seasonal time series of precipitation amount and 10 extreme precipitation indices. Based on climate conditions of Japan, four seasons are divided

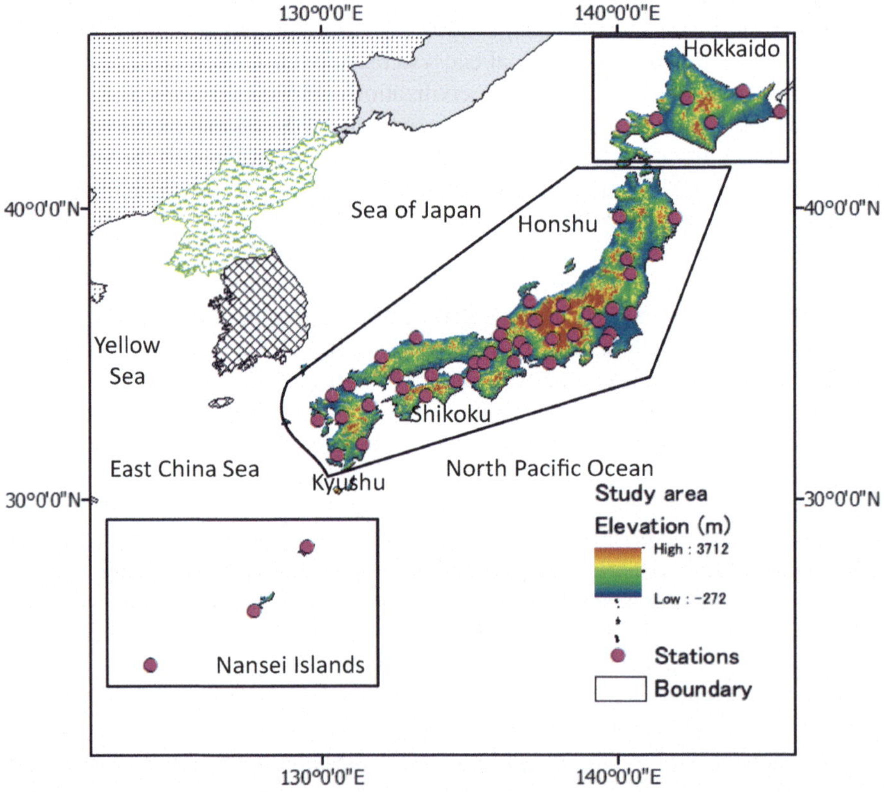

Fig. 5.1 Study area and weather stations

Table 5.1 Weather stations

Station ID	Station number	Station name	Longitude	Latitude	Station ID	Station number	Station name	Longitude	Latitude
J_D1	47409	Abashiri	144.2783	44.01667	J_D27	47629	Mito	140.4667	36.38
J_D2	47582	Akita	140.0983	39.71667	J_D28	47585	Miyako	141.965	39.64667
J_D3	47617	Alpine	137.2533	36.155	J_D29	47830	Miyazaki	131.4133	31.93833
J_D4	47407	Asahikawa	142.3683	43.77167	J_D30	47610	Nagano	138.1917	36.66167
J_D5	47742	Border	133.235	35.54333	J_D31	47817	Nagasaki	129.8667	32.73333
J_D6	47616	Fukui	136.2217	36.055	J_D32	47636	Nagoya	136.965	35.16667
J_D7	47807	Fukuoka	130.375	33.58167	J_D33	47936	Naha	127.685	26.20667
J_D8	47595	Fukushima	140.47	37.75833	J_D34	47909	Naze	129.495	28.37833
J_D9	47606	Fushiki	137.055	36.79167	J_D35	47420	Nemuro	145.585	43.33
J_D10	47632	Gifu	136.7617	35.4	J_D36	47417	Obihiro	143.2117	42.92
J_D11	47755	Hamada	132.07	34.89667	J_D37	47815	Oita	131.6183	33.235
J_D12	47654	Hamamatsu	137.7183	34.70833	J_D38	47772	Osaka	135.5183	34.68167
J_D13	47761	Hikone	136.2433	35.275	J_D39	47412	Sapporo	141.3283	43.05833
J_D14	47637	Iida	137.8217	35.52333	J_D40	47762	Shimonoseki	130.925	33.94833
J_D15	47918	Ishigaki Island	124.1633	24.33667	J_D41	47421	Suttsu	140.2233	42.795
J_D16	47592	Ishinomaki	141.2983	38.42667	J_D42	47890	Tadotsu	133.7517	34.275
J_D17	47827	Kagoshima	130.5467	31.55333	J_D43	47895	Tokushima	134.5733	34.06667

(continued)

Table 5.1 (continued)

Station ID	Station number	Station name	Longitude	Latitude	Station ID	Station number	Station name	Longitude	Latitude
J_D18	47770	Kobe	135.2117	34.69667	J_D44	47662	Tokyo	139.76	35.69
J_D19	47893	Kochi	133.5483	33.56667	J_D45	47651	Tsu	136.52	34.73333
J_D20	47638	Kofu	138.5533	35.66667	J_D46	47631	Tsuruga	136.0617	35.65333
J_D21	47626	Kumagai	139.38	36.15	J_D47	47615	Utsunomiya	139.8683	36.54833
J_D22	47819	Kumamoto	130.7067	32.81333	J_D48	47777	Wakayama	135.1633	34.22833
J_D23	47759	Kyoto	135.7317	35.015	J_D49	47766	Kure	132.55	34.24
J_D24	47624	Maebashi	139.06	36.405	J_D50	47588	Yamagata	140.345	38.255
J_D25	47618	Matsumoto	137.97	36.245	J_D51	47670	Yokohama	139.6517	35.43833
J_D26	47887	Matsuyama	132.7767	33.84333					

as winter = December, January, February; spring = March, April, May; summer = June, July, August; autumn = September, October, November.

In order to identify the spatial relationship between precipitation extreme events and the sea surface temperatures (SSTs), the extended reconstruction sea surface temperature (ERSST.v3b) (Smith et al. 2008) from 1901 to 2012 was used, which has a resolution of 2° latitude × 2° longitude and has been proven to be suitable for long-term global and basin-wide studies.

The computer program RClimDex (Zhang and Yang 2004) was used to evaluate data quality, which can identify potentially unrealistic climatic records, including negative values of daily maximum-minus-minimum temperature outliers (typically exceeding four standard deviations difference from the mean) and negative values of daily precipitation (Alexander et al. 2006; Zongxing et al. 2012). After carrying out the data quality control, we found there were more than three missing or unrealistic climatic records in a month (see Table 5.2). Then, the RHTest software, which was developed by the Climate Research Branch of Meteorological Service of Canada, was employed to determine if there were artificial changes at the station (such as station moves) that significantly impacted the observations (Aguilar et al. 2009). The result of homogeneity checks shows that precipitation data at all stations are reasonable.

Table 5.2 List of stations with more than three missing records

Station ID	Station number	Station name	Periods of unrealistic climatic records	Number of months
J_D6	47616	Fukui	February–December 1938 (except May, July, and October); July–August 1945	11
J_D20	47638	Kofu	June–July 1945	2
J_D45	47651	Tsu	July 1989	1
J_D12	47654	Hamamatsu	June 1945	1
J_D51	47670	Yokohama	August–December 1923	5
J_D49	47766	Kure	April 1945; June 1945–September 1946	17
J_D18	47770	Kobe	March 1945	1
J_D33	47936	Naha	January–July 1923; October 1943: September 1944; February 1945–December 1950; February–March 1951	83

Table 5.3 Definitions of 10 precipitation indices used in this study

ID	Indicator name	Definitions	Units
RX1day	Max 1-day precipitation amount	Monthly maximum 1-day precipitation	mm
RX5day	Max 5-day precipitation amount	Monthly maximum consecutive 5-day precipitation	mm
SDII	Simple daily intensity index	Annual total precipitation divided by the number of wet days (defined as RR >= 1.0 mm) in the year	mm/day
R10 mm	Number of heavy precipitation days	Annual count of days when RR >= 10 mm	day
R20 mm	Number of very heavy precipitation days	Annual count of days when RR >= 20 mm	day
CDD	Consecutive dry days	Maximum number of consecutive days with RR < 1 mm	day
CWD	Consecutive wet days	Maximum number of consecutive days with RR >= 1 mm	day
R95p	Very wet days precipitation	Annual total PRCP when RR > 95th percentile of precipitation on wet days in the 1961–1990 period	mm
R99p	Extremely wet days precipitation	Annual total PRCP when RR > 99th percentile of precipitation on wet days in the 1961–1990 period	mm
PRCPTOT	Annual total wet-day precipitation	Annual total PRCP in wet days (RR >= 1 mm)	mm

Abbreviations are as follows: RR, daily precipitation. A wet day is defined when RR >= 1 mm and a dry day when RR < 1 mm

5.2.2 Selected Extreme Precipitation Indices

Except for precipitation amounts, 10 extreme precipitation indices were calculated to analyze extremes and detect precipitation variations (see Table 5.3). The indices were developed and recommended by the Expert Team on Climate Change Detection and Indices (ETCCDI, available at http://www.climdex.org/indices.html) (Sillmann et al. 2013; Zhang et al. 2011).

5.2.3 Area Averaging and Trend Calculation

All seasonal precipitation and extreme indices and the anomalies of these indices were computed. The base period for anomalous selection is 1981–2010. Positive anomaly values indicate that the precipitation index from 1980 to 2010 is greater than the average precipitation index, while negative anomaly values indicate that the observed precipitation index is less than the average precipitation index from 1980 to 2010.

The Kendall's tau test (Press et al. 1992; Helsel and Hirsch 1992) was used to calculate the trend to determine if statistically significant trends exist in seasonal precipitation amounts as well as in measures of precipitation extremes, through the study time. Regional analysis plays a pivotal role in the changing climate as it can describe, compare, and explore climate changes between different regions and has been used by many studies (Giorgi and Francisco 2000). Therefore, the regional average anomaly sequence for each precipitation index is calculated by the following formula:

$$x_{r,t} = \sum_{i=1}^{n_t} (x_{i,t} - \overline{x_i})/n_t \tag{5.1}$$

where $x_{r,t}$ is the regionally averaged extreme precipitation index at year t, $x_{i,t}$ is the index for weather station i at year t, $\overline{x_i}$ is the 1901–2012 index mean at station i, and n_t is the number of weather stations with data in the year t.

5.2.4 Comparisons to SSTs

In order to determine whether there is a relationship between sea surface temperature (SST) and extreme precipitation events, the Kendall's tau rank correlation between SST and several indices was calculated. Aguilar et al. (2005) used this method to analyze the daily extreme climate in Central and Northern South America.

5.2.5 Atmospheric Circulation Analysis

To better understand the atmospheric circulation patterns behind the spatial and temporal variations of precipitation, water vapor flux and related transport characteristics were explored using NCAR/NCEP reanalysis data (Xu et al. 2004). Due to the availability of the NCAR/NCEP reanalysis data, we calculated changes in water vapor flux and other features annually and seasonally for the periods 1980–1995 and 1996–2011, respectively, which could explain impacts of atmospheric circulation patterns on precipitation in recent decades. According to the European Centre for Medium-Range Weather Forecasts (ECMWF) analyses (Cullather et al. 1998) and others (Zhang et al. 2008, 2011), atmospheric moisture is considered negligible above 300 hPa and so only layers under 300 hPa ($z \leq 8$) were used to calculate the water vapor flux. The zonal moisture transport flux (Q_u), meridional moisture transport flux (Q_v), and whole layer moisture budget (Q_T) at regional boundaries were calculated based on the following equations (Miao et al. 2005; Zhang et al. 2009, 2011):

$$Q_v(x, y, t) = \frac{1}{g} \int_p^{p_s} q(x, y, p, t) * v(x, y, p, t) dp \tag{5.2}$$

$$Q_W = \sum_{\phi_1}^{\phi_2} Q_u(\lambda_1, y, t) \tag{5.3}$$

$$Q_E = \sum_{\phi_1}^{\phi_2} Q_u(\lambda_2, y, t) \tag{5.4}$$

$$Q_S = \sum_{\lambda_1}^{\lambda_2} Q_v(x, \phi_1, t) \tag{5.5}$$

$$Q_N = \sum_{\lambda_1}^{\lambda_2} Q_v(x, \phi_2, t) \tag{5.6}$$

$$Q_T = Q_W - Q_E + Q_S - Q_N \tag{5.7}$$

where u and v are the zonal and meridional components of the wind field, respectively; q is the specific humidity; p_s is surface pressure (z = 1); p is 300 hPa (z = 8); g is acceleration of the gravity; Q_W, Q_E, Q_S, Q_N are the West, East, South, and North regional boundaries, respectively; and ϕ_1, ϕ_2, λ_1, λ_2 are the latitude and longitude according to the regional boundaries (Miao et al. 2005).

5.3 Results and Discussions

5.3.1 Precipitation Amounts and Trends

5.3.1.1 Annual Precipitation Amounts and Trends

Figure 5.2 indicates that precipitation in Japan has varied widely in the past 112 years. Figure 5.2a illustrates the time series of regional average precipitation of Japan, indicating that precipitation during the period from 1901 to 2012 fluctuated year by year. Kendall's tau value was −0.087, suggesting that the observed annual average precipitation has been greatly reduced over the past years. More specifically, the solid trend line shows that annual regional precipitation has decreased by 72.4 mm over the past years. The precipitations in 1923 (about 1925.89 mm) and 1994 (about 1142.44 mm) were the highest and lowest years, respectively, consistent with the wettest year and

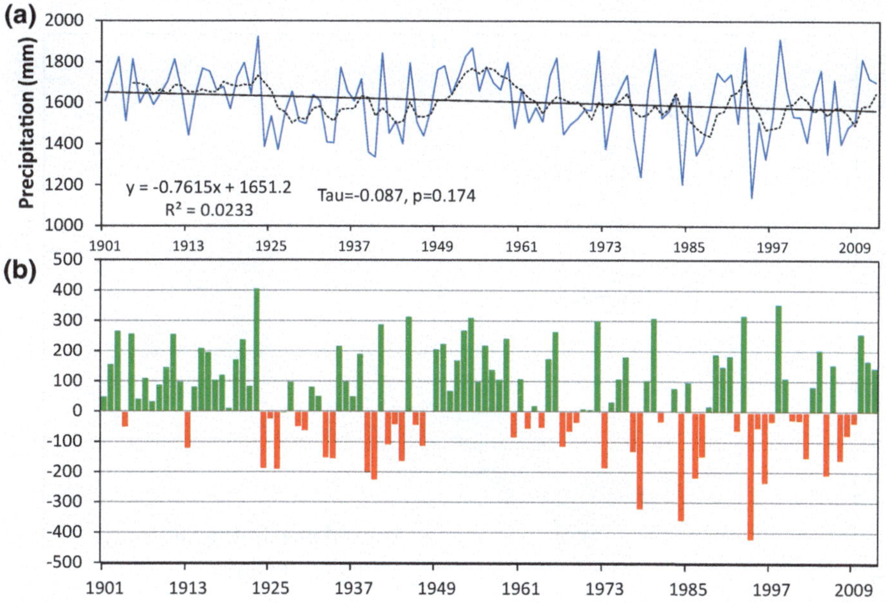

Fig. 5.2 **a** Changes of regionally averaged rainfall amounts (mm) with line trend (straight line) and 9-year running mean (dotted curve) in Japan from 1901 to 2012. **b** National precipitation departures in Japan, 1901–2012, based on the average from 1981 to 2010

the driest year shown in Fig. 5.2b. Compared to Fig. 5.2a, b, precipitation fluctuations are more frequent and intense, especially since the 1960s.

Figure 5.2b shows precipitation departures in Japan, which is based on the average precipitation from 1981 to 2010. These intervals include (1) 8 years with deficits exceeding 200 mm and 22 years with surpluses exceeding 200 mm; (2) two wettest periods (1901–1923 and 1948–1959) and two driest periods (1924–1947 and 1960–2012); (3) the wettest consecutive 12 years (1948–1959) of any 12-year interval; and (4) 1923 is the wettest year since records and 1994 is the driest year since records.

Figure 5.3 shows that the annual mean precipitation ranged from 836 to 2990 mm across Japan. As can be seen from the figure, Hokkaido had less precipitation compared with the other regions and there was much higher precipitation in southwest compared with the northeast. Figure 5.3 also indicates that there were 45 stations with a decreasing trend in precipitation (approximately 88% of the total number of stations), suggesting an overall decline in precipitation from 1901 to 2012. Stations with significance at 95% confidence are mainly distributed in the southeastern Japan.

5.3.1.2 Seasonal Precipitation Amounts and Trends

Like the annual data results, more than 57% of stations in all four seasons tended to decrease (see Table 5.4); in winter, the number of negative stations was the highest

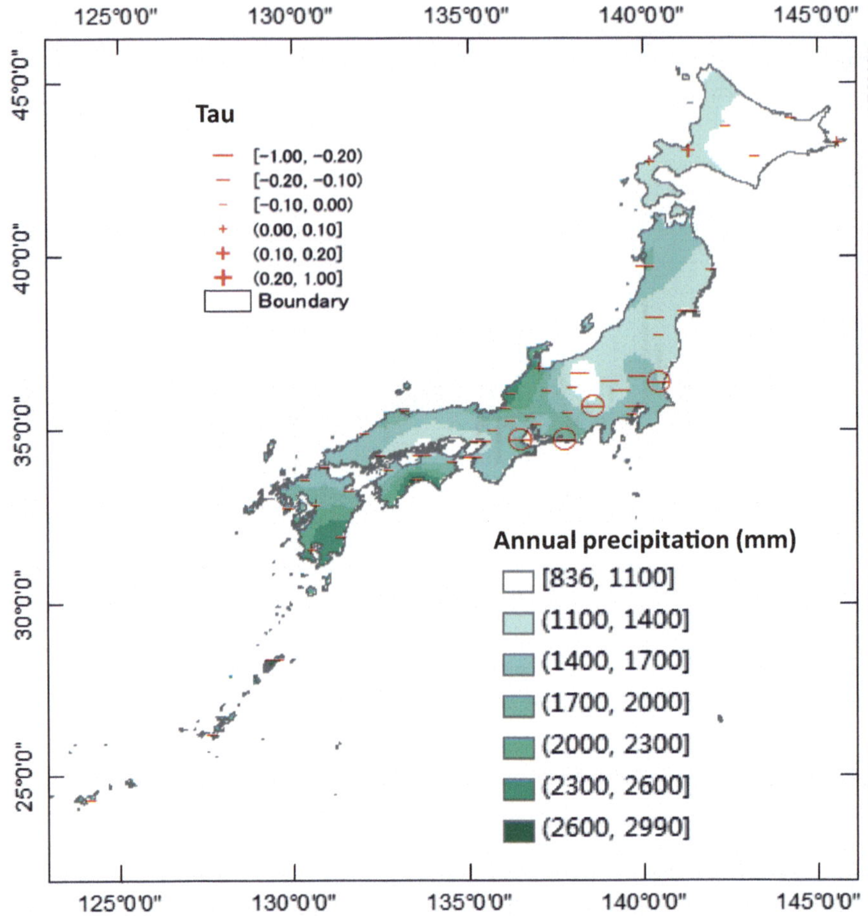

Fig. 5.3 Annual mean precipitation (mm), trends (Kendall's tau) for 51 stations and changes of regionally averaged rainfall amounts (mm) in Japan from 1901 to 2012. Positive trends are shown as pluses and negative trends as minuses. Trends that are significant at the 95% level are circled

(49, about 96%), followed by autumn (46, about 90%), summer (34, about 67%), and spring (29, about 57%). Among them, there were 11 stations in autumn (about 22%, 11 positive and 0 negative) significance at 95% confidence, while it fell to 5 (4 positive and 1 negative) in winter, 2 stations (1 positive and 1 negative) in summer, and 0 station in spring.

Like the annual trend, seasonal precipitation over a hundred-year time series shows a slight (and negligible) downward trend (Fig. 5.4). The total precipitation in autumn decreased the most value among four seasons (Tau = −0.116, p = 0.071), mainly due to negative anomalies since the early 1960s (Fig. 5.4c). The second largest decline in winter (Tau = −0.115, p = 0.072) was mainly due to negative anomalies since the early 1970s (Fig. 5.4d); spring (Tau = −0.009, p = 0.890, Fig. 5.4a) and summer

Table 5.4 Annual trends and percentage of stations with positive or negative trends for regional indices of precipitation extremes in Japan during 1901–2012

ID	Regional trends	Range	Showing positive trend	Showing significant positive trend	Showing negative trend	Showing significant negative trend
RX1day	0.114	−0.134–0.229	37	2	14	1
RX5day	0.115	−0.107–0.17	38	5	13	0
SDII	**0.140**	−0.136–0.244	43	9	8	1
R10 mm	**−0.129**	−0.208–0.169	7	1	44	13
R20 mm	−0.023	−0.176–0.164	16	1	35	1
CDD	**0.237**	−0.0377–0.223	47	23	4	0
CWD	**−0.178**	−0.236–0.249	11	2	40	22
R95p	0.083	−0.0856–0.192	38	3	13	0
R99p	**0.150**	−0.142–0.297	40	6	11	1
PRCPTOT	−0.081	−0.161–0.133	7	1	44	4
Annual precipitation	−0.087	−0.169–0.117	6	0	45	4
Spring precipitation	−0.009	−0.124–0.0759	23	0	29	0
Summer precipitation	−0.005	−0.188–0.128	27	1	34	1
Autumn precipitation	−0.116	−0.169–0.0795	5	0	46	11
Winter precipitation	−0.115	−0.222–0.175	2	1	49	4

(Tau = −0.005, p = 0.934, Fig. 5.4b) were negligible. Therefore, the decrease in total precipitation in autumn and winter was the main reason for the decrease in total annual precipitation (Fig. 5.2). Figure 5.4 also shows the wettest year period and high positive and negative anomalies in different seasons. The summer had the largest positive anomaly and negative deficit (326 mm in 1905 and −252 mm in 1994), followed by autumn (301 mm in 1945 and −179 mm in 1984), spring (154 mm in 1903 and −116 mm in 2005), and winter (129 mm in 1915 and −69 mm in 1996).

The seasonal differences in precipitation distribution are shown in Fig. 5.5, which suggests that summer and autumn had the high precipitation. In addition, in spring, summer, and autumn, the precipitation in southwestern Japan was higher than that in the northeast. In winter, the precipitation in the northwest was higher than that in the northeast, indicating that the spatial distribution in Japan was uneven in the past 112 years. Figure 5.5 also shows the spatial distribution of trends based on 51 stations. In spring and summer, precipitation in the northwest increased, while precipitation in the southeast decreased. In the winter, some stations with significantly negative

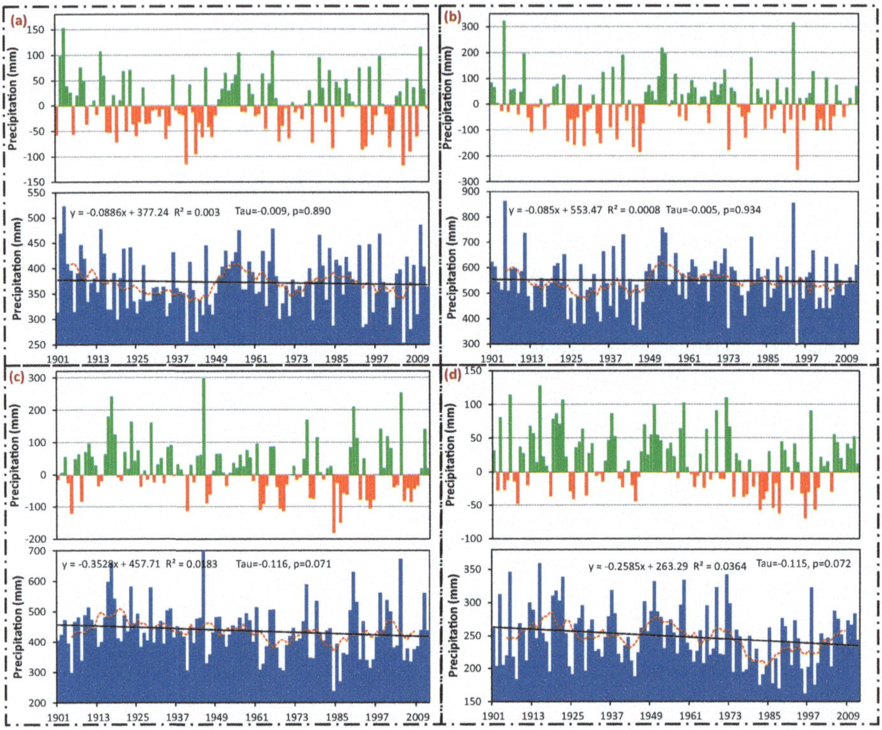

Fig. 5.4 As in Fig. 5.2 but for four seasons: **a** Spring; **b** Summer; **c** Autumn; **d** Winter

trends were mostly scattered on the northern edge. In contrast, in the autumn, the sporadic stations in the southeast showed a significant negative trend. These seasonal changes are also consistent with the results from Kimoto et al. (2005) and Fujibe et al. (2005).

5.3.1.3 Changes of Annual Precipitation Extremes

In the past 112 years, some precipitation extremes of most stations, such as R10 mm, R20 mm, CWD, and PRCPTOT, have shown a downward trend (Table 5.4). Among them, R10 mm and PRCPTOT stations had the largest number, with a negative trend (44 stations, about 86%), followed by CWD (40 stations, about 78%) and R20 mm (35 stations, about 69%). On the other hand, for most other precipitation extremes, such as R95p, R99p, CDD, RX1day, RX5day, and SDII, it has been found that there was an increasing trend in most stations (Table 5.4). Of these sites, almost all sites showed a positive trend for CDD (47 sites), followed by SDII (43 sites, approximately 84%), R99p (40 sites, approximately 78%), R95p (38 sites, approximately 75%), RX 5 days (38 stations, approximately 75%), and RX1day (37 stations, approximately 73%). For regional average trends, changes in R10 mm, R20 mm, CWD, and PRCPTOT

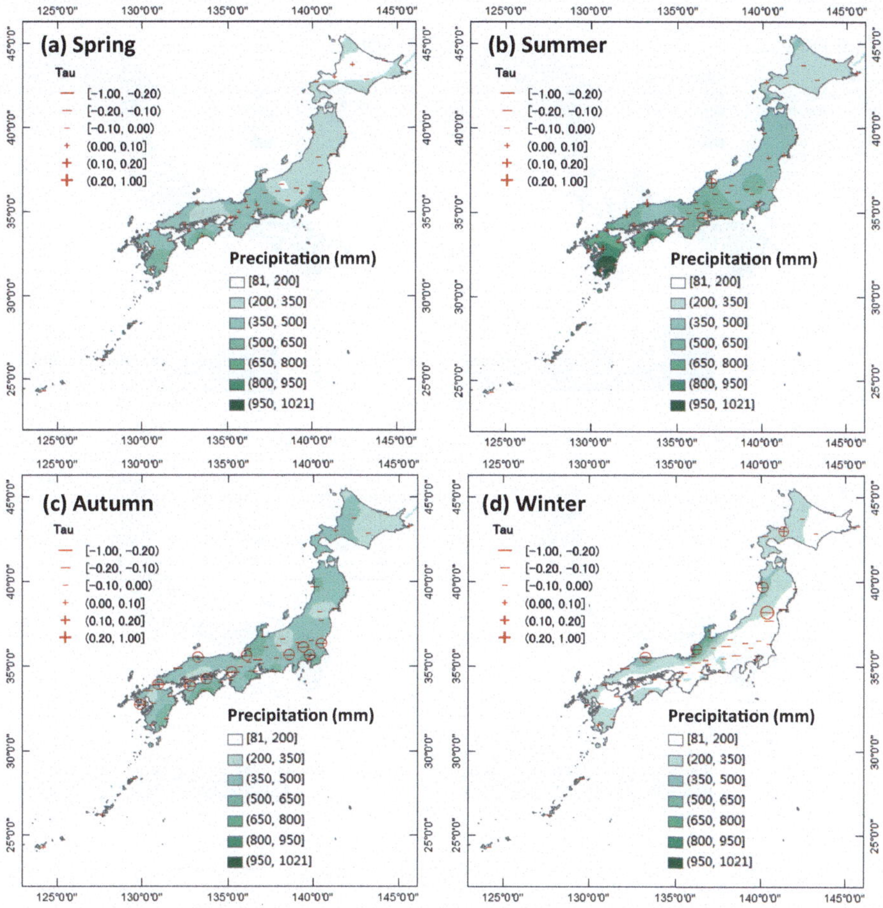

Fig. 5.5 As in Fig. 5.4 but for four seasons: **a** Spring; **b** Summer; **c** Autumn; **d** Winter

have also declined over the past 112 years (Table 5.4), while R95p, R99p, CDD, RX1day, RX5day, and SDII were on the rise.

Like the total precipitation, the precipitation extremes in southwestern Japan were usually higher than in the northeast (Fig. 5.6). For example, the SDII in the southwest ranged from 13 mm day^{-1} to 18 mm day^{-1}, while the SDII in the northeast ranged from 7 mm day^{-1} to 13 mm day^{-1}. More precisely, Hokkaido's extreme index value was much lower than in other parts of Japan.

Except for Hokkaido, the negative trends of PRCPTOT, R10 mm, and R20 mm were dominant, and the stations with statistically significant trends were mainly distributed in the southeastern part of Japan. In contrast, extremely wet daily precipitation (R95p), extremely wet daily precipitation (R99p), and average wet daily precipitation (SDII) had generally increased, and stations with statistically significant trends were mainly distributed in southwestern Japan. In addition, although the total

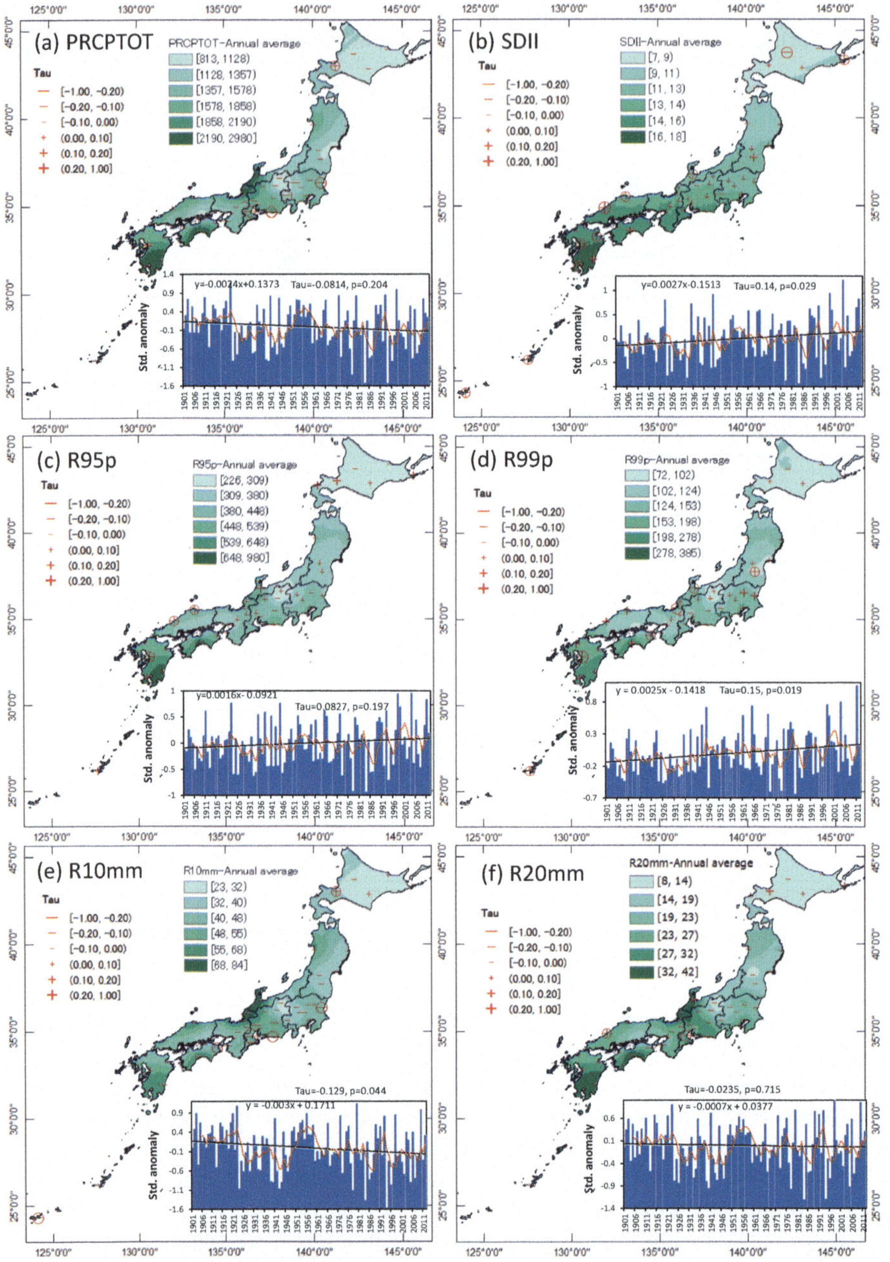

Fig. 5.6 Spatial pattern of trends (Kendall's tau), spatial distribution of annual mean, and regional averaged standardized series for precipitation extremes indices. Positive trends are shown as pluses and negative trends as minuses. Trends that are significant at the 95% level are circled. Insets show the regionally averaged standardized anomalies relative to 1981–2010

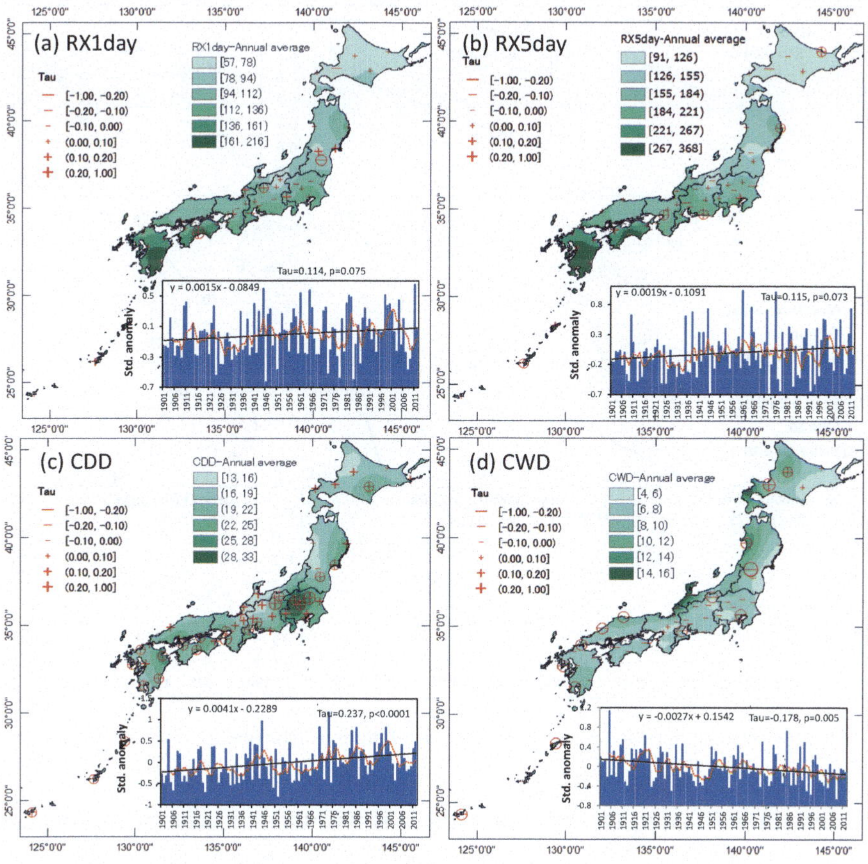

Fig. 5.7 As in Fig. 5.6 but for precipitation spell indices

precipitation had decreased over the past 112 years (Figs. 5.2 and 5.3), the maximum 1-day precipitation and the maximum 5-day precipitation had increased (Fig. 5.7a, b). As shown in Fig. 5.7c, for continuous dry days (CDD), most of the weather stations with positive trends were observed, suggesting the number of consecutive dry days increased over the past years. Conversely, the number of consecutive wet days had decreased over the past 112 years, which largely indicates that most negative trend stations were scattered throughout Japan except for the Hokkaido region (Fig. 5.7d).

5.3.2 *Relations with SSTs*

Sea surface temperatures influence precipitation patterns throughout the world, so when sea surface temperatures change, precipitation patterns tend to change as well. Figure 5.8 shows that the averaged anomalies of surface temperatures in global and

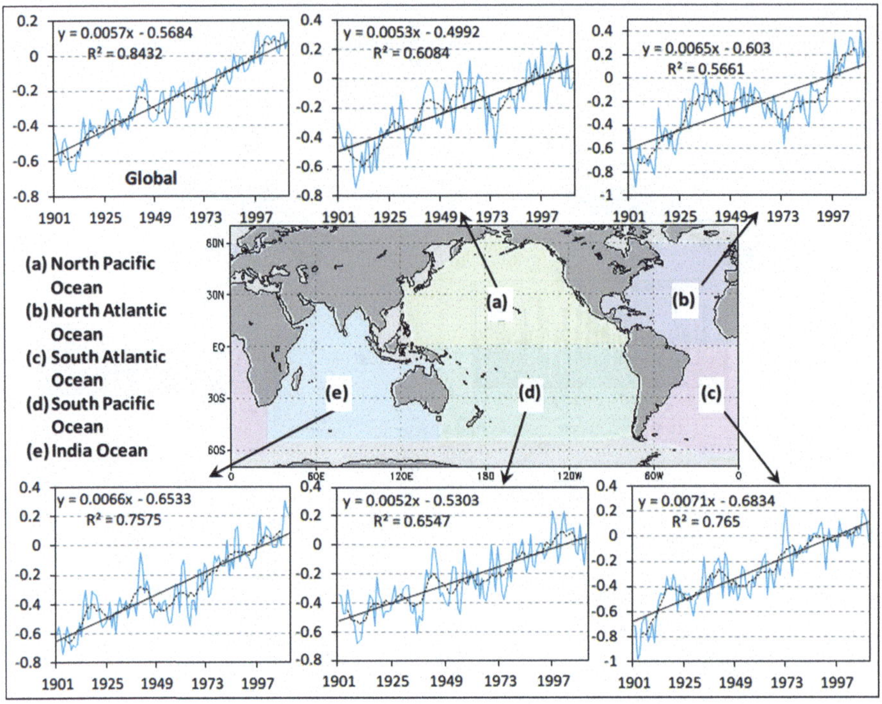

Fig. 5.8 Regional averaged anomalies of surface temperatures in main oceans relative to 1981–2010

in main oceans including North Pacific Ocean, North Atlantic Ocean, South Atlantic Ocean, South Pacific Ocean, and Indian Ocean, relative to 1981–2010, indicating the averaged SSTs increased over the twentieth century and will probably continue to rise. Among these, the North Pacific Ocean plays an important role in Japan's precipitation.

Figure 5.9 gives the correlation patterns between precipitation extreme indices and SSTs by calculating the Kendall's tau, which indicates that generally the precipitation extreme indices had a positive correlation with the Pacific, the Sea of Japan, and the China East Sea, with the exception of PRCPTOT. Moreover, the positive correlation was generally stronger in the sea areas around Japan. Except for some offshore areas, as shown in Fig. 5.9a, the correlation was negative in most sea areas, suggesting that the total wet-day precipitation decreased over the past 112 years with the SSTs increasing.

The SSTs variation is a key factor for the precipitation changes, because increased evaporation of a warmer ocean not only lead to more water vapor available for precipitation but also to the potential relationship with a longer or more intense tropical cyclone (Aguilar et al. 2005). Increases in SSTs generally bring more precipitation, but actually the annual total wet-day precipitation (PRCPTOT) decreased in Japan over the past century. Why? Japan's climate is characterized as monsoonal (i.e.,

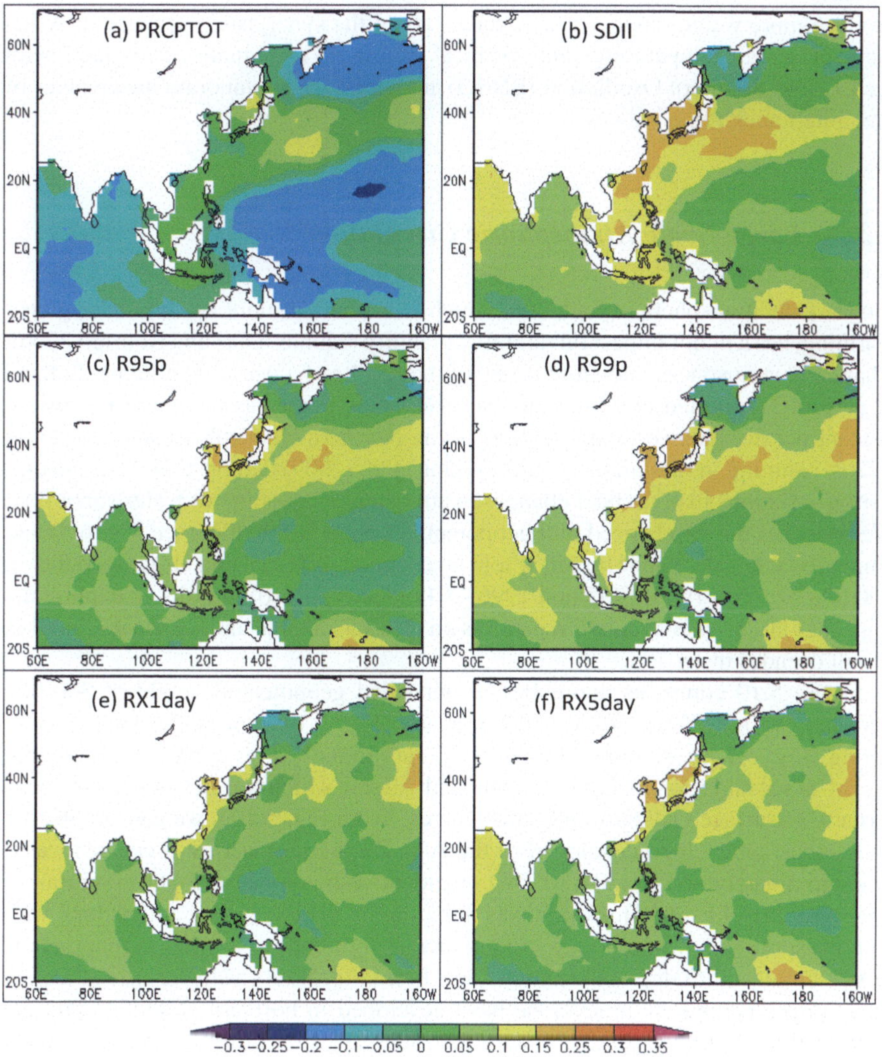

Fig. 5.9 Correlation patterns of annual precipitation extreme indices and sea surface temperatures from 1901 to 2012. **a** PRCPTOT; **b** SDII; **c** R95p; **d** R99p; **e** RX1day; **f** RX5day

governed by seasonal winds), that is, wet weather is brought by East Asian summer monsoon from the Pacific Ocean and Southeast Asia in summer, and freezing temperatures and heavy snowfalls always happen caused by the winter monsoon from late September to late March. However, the weakening of the East Asian summer monsoon and Asian winter monsoon has been observed in many recent studies (Li et al. 2010; Nakamura et al. 2002; Panagiotopoulos et al. 2005). It may explain the downtrend in annual total wet-day precipitation, although SSTs increasing may produce more water vapor. Further studies about the synergistic influence of Asian land surface air temperatures and Tropical Pacific SST anomalies associated with the El Niño-Southern Oscillation (ENSO) are necessary to elaborate the changes of precipitation extreme indices.

5.3.3 Large-Scale Atmospheric Circulation

Large-scale atmospheric circulation, which is becoming more erratic under global warming, has a major impact on rainfall variability (Muñoz-Díaz and Rodrigo 2006). Change in atmospheric moisture is one of the results of a warming climate. Folland et al. (2002) reported that total atmospheric water vapor has increased by several percent per decade over many regions of the Northern Hemisphere since the early 1970s, resulting in rainfall variability; Zhang et al. (2009) demonstrated that precipitation variations likely reflect changes in atmospheric moisture or water vapor flux. We analyzed the wind speed and geopotential height at 500 hPa (see Fig. 6.6) and compared the water vapor flux in whole layers under 300 hPa and air temperature at 500 hPa between 1980–1995 and 1996–2011 (see Fig. 5.10) in order to explore the impact of large-scale circulation patterns on trends in annual and seasonal rainfall in recent decades in Japan.

Figure 5.10 compares anomalies in wind and geopotential height at 500 hPa between 1980–1995 and 1996–2011 annually (Fig. 5.10a), in winter (Fig. 5.10b), spring (Fig. 5.10c), summer (Fig. 5.10d), and autumn (Fig. 5.10e). Geopotential height is valuable for locating troughs and ridges, which are the upper air counterparts of surface cyclones and anticyclones. From Fig. 5.10a, we can see that an anomalous anticyclone developed in the Asia-Pacific region with a center located at approximately 46° N and 180° W coincident with a negative trend in rainfall in the north of Hokkaido. Hokkaido was therefore on the edge of an anticyclone with southwesterly wind, which may explain increasing trends in rainfall in that there is more rainfall at the edge of the anticyclone. From Fig. 5.10b, as in the case of annual data, a cyclone and an anticyclone have developed in northern Hokkaido and the Asia-Pacific region, respectively, but the centers are a bit different. An anomalous cyclone, developed in Japan near 37° N and 140° E, can be seen in Fig. 5.10c, which impacted on the whole of Japan in spring during 1980–1995 as compared to 1996–2011. This may be a reason why spring rainfall is concentrated in the southeast and most of the stations revealed strongly increasing trends in spring, especially in May. Figure 5.10d shows two anticyclones on both sides of Hokkaido and a small cyclone in northern Hokkaido, which may explain why there are a few stations showing negative trends in the southeast in summer. Figure 5.10e also shows that an outstanding cyclonic circulation anomaly is located to the south of Lake Baikal and an anomalous

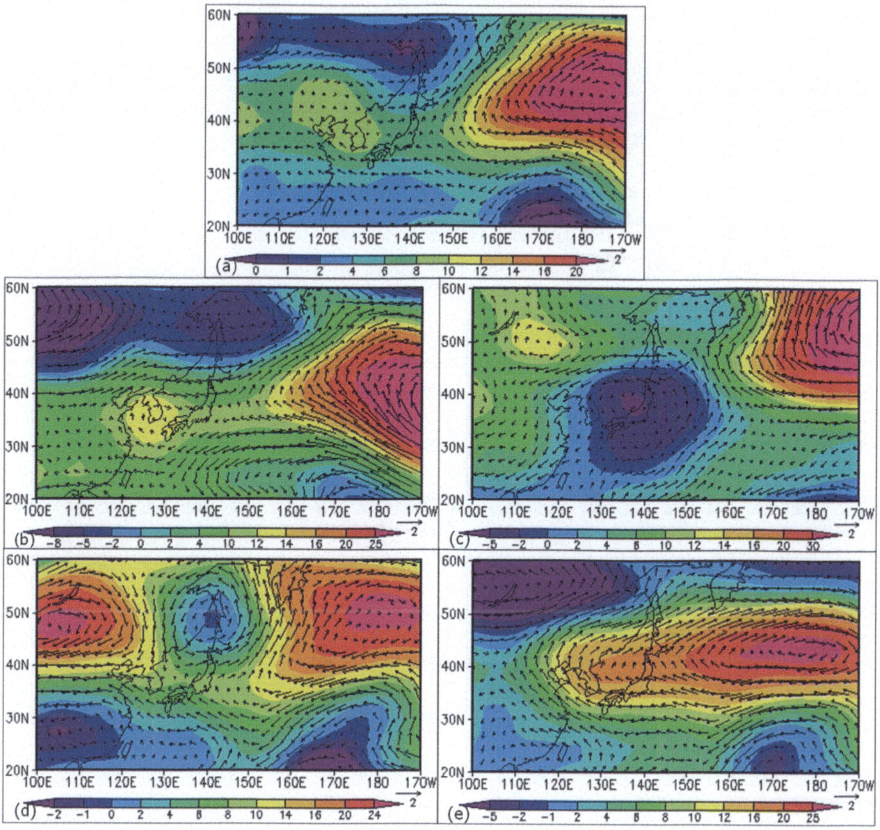

Fig. 5.10 Difference of wind speed (m/s) and geopotential height (gpdm) at 500 hPa in annual (**a**), winter (**b**), spring (**c**), summer (**d**), and autumn (**e**), between 1980–1995 and 1996–2011

anticyclone is located to the west of the Sea of Okhotsk, forming a northwest–southeast pressure gradient around Japan, which may explain higher rainfall values in the northwest compared to the southeast in autumn.

Figure 5.11 shows that the changes in air temperature at 500 hPa in Japan show increasing trends between 1980–1995 and 1996–2011, with autumn showing the biggest change of almost 0.7 °C. These results are in line with the results indicated by You et al. (2011), showing an increasing trend at 500 hPa. Because a warmer atmosphere holds more water vapor (Cline 1991), temperature increases are likely to impact water vapor flux and therefore contribute to rainfall variability. Figure 5.11a clearly shows that there were increasing trends in water vapor flux between 1980–1995 and 1996–2011 all over Japan, which may explain increasing rainfall in these years. On a seasonal timescale consistent with overall increases in rainfall, increases in water vapor flux were mainly seen in winter (Fig. 5.11b), spring (Fig. 5.11c),

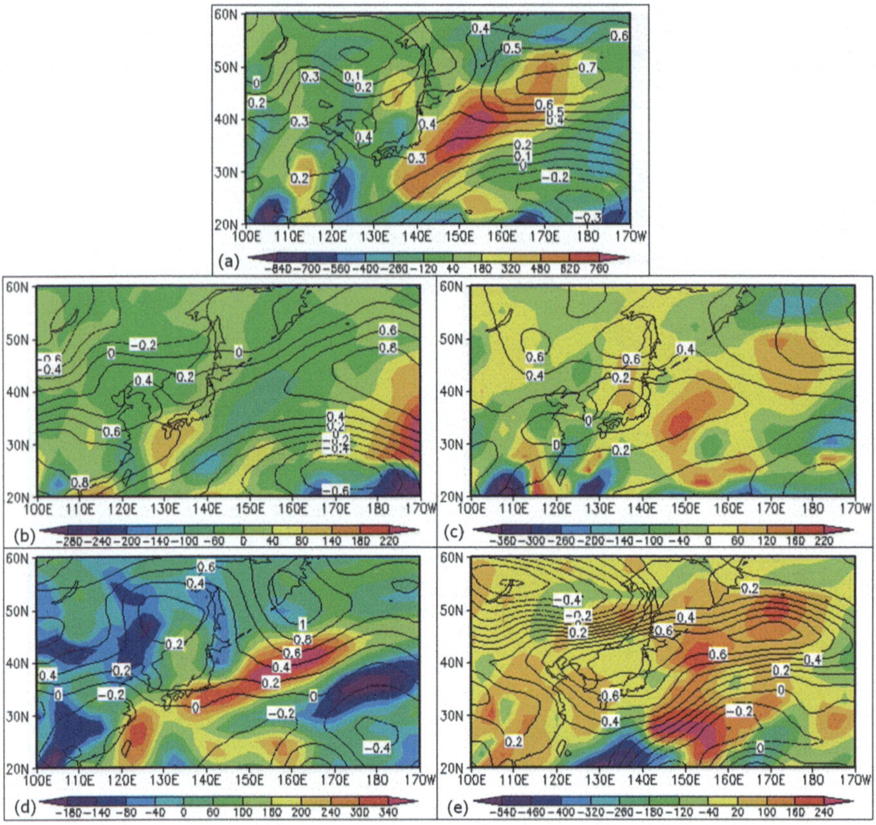

Fig. 5.11 Difference of water vapor flux (kg/m s) in whole lays under 300 hpa and air temperature (Celsius degree) at 500 hPa in annual (**a**), winter (**b**), summer (**c**), autumn (**d**), and spring (**e**), between 1980–1995 and 1996–2011

and autumn (Fig. 5.11e). However, decreases were seen in the southeast in summer (Fig. 5.11d), which does not fit well with the above change observation. In general, these changes are in line with former researchers. Kurihara et al. (2005) pointed out that changes of surface temperatures in and around Hokkaido were predicted to be larger than those in other regions of Japan, and indicated that anticyclonic circulation in the south of Japan will likely intensify and induce a strong water vapor flux along the rim of the anticyclonic anomaly in the future, which would increase the likelihood of heavy rainfall. Westerly moisture flux associated with the prominent pattern of the surface pressure anomalies between the main island of Japan and the Sea of Okhotsk also influences the rainfall in Japan (Yamada et al. 2012).

5.4 Conclusions

Major conclusions can be summarized as follows:

(1) Precipitation amounts exhibited a substantial decrease at both the annual and seasonal scales, with 49 stations (approximately 96% of the total number of stations) showing negative trends on a seasonal basis in winter, followed by autumn (46, approximately 90%), annual basis (45, approximately 88%), summer (34, approximately 67%) and spring (29, approximately 57%). The fluctuation of precipitation became more frequent and intense at both the annual and seasonal scales in the latest decades.
(2) Precipitation varied substantially in spatial–temporal. Precipitation was mainly concentrated in summer (ranging from 286 mm to 1021 mm) and autumn (ranging from 218 mm to 719 mm), while precipitation amounts were less in winter and spring. The southwest had higher precipitation than the southeast in spring, summer, autumn, and interannually, with precipitation concentrated in the southeast in winter.
(3) Variations in R10 mm, R20 mm, CWD, and PRCPTOT indicated a decreasing trend for a whole Japan, with while an increasing trend for R95p, R99p, CDD, RX1day, RX5day, and SDII. The spatial differences of these indices were obvious. Negative trends dominated for PRCPTOT, R10 mm, and R20 mm, with the exception of the Hokkaido, and stations with statistical significant trends for R95p, R99p, and SDII mainly scattered in the southwest area of Japan.
(4) Except for PRCPTOT, other precipitation extreme indices including SDII, R95p, R99p, RX1day, and RX5day had a positive correlation with the SSTs.
(5) The analyses of large-scale atmospheric circulation using wind speed and geopotential height at 500 hPa and the water vapor flux in whole layers under 300 hpa and air temperature at 500 hPa between 1980–1995 and 1996–2011 could explain changes of the spatiotemporal distributions in precipitation in recent decades in Japan. Due to the changes in air temperature, wind speed, and geopotential height, the resulting water vapor flux increased between 1980–1995 and 1996–2011 across Japan, leading to more rainfall in recent decades in northeast Japan.

References

Aguilar, E., Aziz Barry, A., Brunet, M., Ekang, L., Fernandes, A., Massoukina, M., et al. (2009). Changes in temperature and precipitation extremes in western central Africa, Guinea Conakry, and Zimbabwe, 1955–2006. *Journal of Geophysical Research: Atmospheres, 1984–2012*(114), 10–1029.

Aguilar, E., Peterson, T. C., Obando, P. R., Frutos, R., Retana, J. A., Solera, M., et al. (2005). Changes in precipitation and temperature extremes in Central America and northern South America, 1961–2003. *Journal of Geophysical Research: Atmospheres, 1984–2012*(110), 10–1029.

Alexander, L. V., Zhang, X., Peterson, T. C., Caesar, J., Gleason, B., Klein Tank, A., et al. (2006). Global observed changes in daily climate extremes of temperature and precipitation. *Journal of Geophysical Research: Atmospheres, 1984–2012*(111), 10–1029.

Allan, R. P., & Soden, B. J. (2008). Atmospheric warming and the amplification of precipitation extremes. *Science, 321,* 1481–1484.

Burns, A., Gleadow, R., Cliff, J., Zacarias, A., & Cavagnaro, T. (2010). Cassava: The drought, war and famine crop in a changing world. *Sustainability, 2,* 3572–3607.

Cline, W. R. (1991). Scientific basis for the greenhouse effect. *The Economic Journal, 101,* 904–919.

Coumou, D., & Rahmstorf, S. (2012). A decade of weather extremes. *Nature Climate Change, 2,* 491–496.

Cullather, R. I., Bromwich, D. H., & Van Woert, M. L. (1998). Spatial and temporal variability of antarctic precipitation from atmospheric methods. *Journal of Climate, 11,* 334–367.

Dai, A. (2011). Drought under global warming: A review. *Wiley Interdisciplinary Reviews: Climate Change, 2,* 45–65.

Duan, W., Hanasaki, N., Shiogama, H., Chen, Y., Zou, S., Nover, D., et al. (2019). Evaluation and future projection of chinese precipitation extremes using large ensemble high-resolution climate simulations. *Journal of Climate, 32,* 2169–2183.

Duan, W., He, B., Takara, K., Luo, P., Hu, M., Alias, N. E., et al. (2015). Changes of precipitation amounts and extremes over Japan between 1901 and 2012 and their connection to climate indices. *Climate Dynamics, 45,* 2273–2292.

Easterling, D. R., Meehl, G. A., Parmesan, C., Changnon, S. A., Karl, T. R., & Mearns, L. O. (2000). Climate extremes: Observations, modeling, and impacts. *Science, 289,* 2068–2074.

Folland, C. K., Karl, T. R., & Jim Salinger, M. (2002). Observed climate variability and change. *Weather, 57,* 269–278.

Fujibe, F., Yamazaki, N., Katsuyama, M., & Kobayashi, K. (2005). The increasing trend of intense precipitation in Japan based on four-hourly data for a hundred years. *SOLA, 1,* 41–44.

Fujibe, F., Yamazaki, N., & Kobayashi, K. (2006). Long-term changes of heavy precipitation and dry weather in Japan (1901–2004). *Journal of the Meteorological Society of Japan, 84,* 1033–1046.

Giorgi, F., & Francisco, R. (2000). Evaluating uncertainties in the prediction of regional climate change. *Geophysical Research Letters, 27,* 1295–1298.

Helsel, D. R., & Hirsch, R. M. (1992). *Statistical methods in water resources, studies in environmental science* (Vol. 49, p. 522). New York: Elsevier.

Hu, M., Takara, K., Duan, W., He, B., & Luo, P. (2015). Integrated assessment of hydro-climatology variability in Kamo River Basin: confronting climate and extremes. *International Journal of Sustainable Future for Human Security, 3,* 46–55.

Iwasaki, H., & Sunaga, Y. (2009). Study of recent variation in weak rainfall over Japan using 31-year AMeDAS dataset. *Scientific Online Letters on the Atmosphere, 5,* 157–159.

Kawamoto, N., Oki, R., & Shimizu, S. (2011). Comparison between TRMM/PR and AMeDAS ground rain gauge network in terms of annual rainfall, IEEE, pp. 2590–2593.

Kerr, R. A. (2007). Global warming is changing the world. *Science, 316,* 188–190.

Kimoto, M., Yasutomi, N., Yokoyama, C., & Emori, S. (2005). Projected changes in precipitation characteristics around Japan under the global warming. *Scientific Online Letters on the Atmosphere, 1,* 85–88.

Kurihara, K., Ishihara, K., Sasaki, H., Fukuyama, Y., Saitou, H., Takayabu, I., et al. (2005). Projection of climatic change over Japan due to global warming by high-resolution regional climate model in MRI. *SOLA, 1,* 97–100.

Lenderink, G., & Van Meijgaard, E. (2008). Increase in hourly precipitation extremes beyond expectations from temperature changes. *Nature Geoscience, 1,* 511–514.

Li, H., Dai, A., Zhou, T., & Lu, J. (2010). Responses of East Asian summer monsoon to historical SST and atmospheric forcing during 1950–2000. *Climate Dynamics, 34,* 501–514.

Marengo, J. A., Nobre, C. A., Tomasella, J., Cardoso, M. F., & Oyama, M. D. (2008). Hydro-climatic and ecological behaviour of the drought of Amazonia in 2005. *Philosophical Transactions of the Royal Society B: Biological Sciences, 363,* 1773–1778.

Medina-Elizalde, M., & Rohling, E. J. (2012). Collapse of classic maya civilization related to modest reduction in precipitation. *Science, 335,* 956–959.

Miao, Q. J., Xu, X. D., & Zhang, S. Y. (2005). Whole layer water vapor budget of Yangtze River valley and moisture flux components transform in the key areas of the plateau. *Acta Meteorological Sinica, 63,* 93–99.

Miyajima, J., & Fujibe, F. (2011). Climatology of extreme precipitation in Japan for different time scales. *SOLA, 7,* 157–160.

Muñoz-Díaz, D., & Rodrigo, F. S. (2006). Seasonal rainfall variations in Spain (1912–2000) and their links to atmospheric circulation. *Atmospheric Research, 81,* 94–110.

Nakamura, H., Izumi, T., & Sampe, T. (2002). Interannual and decadal modulations recently observed in the Pacific storm track activity and East Asian winter monsoon. *Journal of Climate, 15,* 1855–1874.

Panagiotopoulos, F., Shahgedanova, M., Hannachi, A., & Stephenson, D. B. (2005). Observed trends and teleconnections of the Siberian high: A recently declining center of action. *Journal of Climate, 18,* 1411–1422.

Press W. H., Flannery B. P., Teukolsky S. A., & Vetterling W. T. (1992). *Numerical recipes, the art of scientific computing*. Cambridge University Press, Cambridge, 2nd edition.

Sillmann, J., Kharin, V. V., Zhang, X., Zwiers, F. W., & Bronaugh, D. (2013). Climate extremes indices in the CMIP5 multimodel ensemble: Part 1. Model evaluation in the present climate. *Journal of Geophysical Research: Atmospheres, 118,* 1716–1733.

Smith, T. M., Reynolds, R. W., Peterson, T. C., & Lawrimore, J. (2008). Improvements to NOAA's historical merged land-ocean surface temperature analysis (1880–2006). *Journal of Climate, 21,* 2283–2296.

Solomon, S. (2007). Climate change 2007: *The physical science basis: contribution of Working Group I to the Fourth Assessment Report of the Intergovernmental Panel on Climate Change*, Cambridge University Press.

Takeshita, S. (2010). Influence on precipitation and the distribution in Miyazaki [Japan] Prefecture with climate changes. *Bulletin of the Faculty of Agriculture, Miyazaki University (in Japanese), 56,* 73–78.

Xu, X., Chen, L., Wang, X., Miao, Q., & Tao, S. (2004). Moisture transport source/sink structure of the Meiyu rain belt along the Yangtze River valley. *Chinese Science Bulletin, 49,* 181–188.

Yamada, T. J., Sasaki, J., & Matsuoka, N. (2012). Climatology of line-shaped rainbands over northern Japan in boreal summer between 1990 and 2010. *Atmospheric Science Letters, 13,* 133–138.

You, Q., Kang, S., Aguilar, E., Pepin, N., Flügel, W. A., Yan, Y., et al. (2011). Changes in daily climate extremes in China and their connection to the large scale atmospheric circulation during 1961–2003. *Climate Dynamics, 36,* 2399–2417.

Zhang, Q., Xu, C., Gemmer, M., Chen, Y. D., & Liu, C. (2009a). Changing properties of precipitation concentration in the Pearl River basin, China. *Stochastic Environmental Research and Risk Assessment, 23,* 377–385.

Zhang, Z., Zhang, Q., Xu, C., Liu, C., & Jiang, T. (2009b). Atmospheric moisture budget and floods in the Yangtze River basin, China. *Theoretical and Applied Climatology, 95,* 331–340.

Zhang, Q., Xu, C. Y., Chen, X., & Zhang, Z. (2011a). Statistical behaviours of precipitation regimes in China and their links with atmospheric circulation 1960–2005. *International Journal of Climatology, 31,* 1665–1678.

Zhang, X., Alexander, L., Hegerl, G. C., Jones, P., Tank, A. K., Peterson, T. C., et al. (2011b). Indices for monitoring changes in extremes based on daily temperature and precipitation data. *Wiley Interdisciplinary Reviews: Climate Change, 2,* 851–870.

Zhang, Q., Xu, C. Y., Zhang, Z., Chen, Y. D., Liu, C., & Lin, H. (2008). Spatial and temporal variability of precipitation maxima during 1960–2005 in the Yangtze River basin and possible association with large-scale circulation. *Journal of Hydrology, 353,* 215–227.

Zhang, X., Yang, F., 2004. RClimDex (1.0) user manual. Climate Research Branch Environment Canada, Downsview.

Zongxing, L., He, Y., Wang, P., Theakstone, W. H., An, W., Wang, X., et al. (2012). Changes of daily climate extremes in southwestern China during 1961–2008. *Global and Planetary Change, 80,* 255–272.

Chapter 6
Precipitation Changes in Hokkaido and Future Water Resources in Its Main Rivers

6.1 Introduction

Continued global warming has severely affected the spatiotemporal distribution of water resources and extreme climate events (Brown and Funk 2008; Coumou and Rahmstorf 2012; Hendrix and Salehyan 2012). For example, Oki and Kanae (2006) argued that the spatiotemporal of precipitation is very uneven, resulting in significant time-varying changes in water resources worldwide. Zhang et al. (2007) found that precipitation tends to increase in the mid-latitudes of the northern hemisphere, drying in the subtropical and tropical regions of the northern hemisphere, and moistening in the subtropical and deep-tropical regions of the southern hemisphere. Evaporation rates vary widely depending on factors such as cloudiness, air temperature and wind speed, and greatly affect the amount of water that can be used to recharge groundwater. Shorter duration but greater rainfall (meaning more runoff and less infiltration) will lead to depletion of groundwater (Konikow and Kendy 2005; Wada et al. 2010).

There is much evidence of the impact of climate change on the hydrological climate in Japan (Solomon 2007). According to a report from the Japan Meteorological Agency (JMA), the annual average temperature rise rate was equivalent to 1.15 °C per century between 1898 and 2010, which was much higher than the global average temperature increase of 0.74 °C in the last century (according to the Intergovernmental Panel on Climate Change's "Climate Change 2007: Synthesis Report Summary for Policymakers"). The Special Committee's "Climate Change 2007: Summary of Policymakers' Comprehensive Reports"; in addition, Japan's annual precipitation varied widely over the past years. All these changes have greatly affected Japan's water supply. Therefore, many authors have analyzed precipitation changes at different time scales in Japan (Iwasaki and Sunaga 2009). Although all of these studies have detected the characteristics of rainfall changes, most studies have focused on interannual or summer variability, and seldom conduct feature analysis on multiple time scales (e.g., year, season, and month). Moreover, based on general circulation models (GCMs)'s output, many efforts have been made to assess Japan's hydrological climate under climate change. For example, Sato et al. (2013) studied the effects

W. Duan and K. Takara, *Impacts of Climate and Human Activities on Water Resources and Quality*, https://doi.org/10.1007/978-981-13-9394-5_6

of climate change on river flows in several major river basins in Japan. However, in the small river basin of Hokkaido, there are few studies on the hydro-climatology variations under climate change.

Therefore, the objective of this chapter is firstly to elucidate precipitation changes at multiple time scales in Hokkaido, using the Moran's I (a measure of spatial auto-correlation), the Mann–Kendall test (a non-parametric trend test) and geostatistical interpolation techniques, and then to investigate the possible effects of climate change on water resources in the upper Ishikari River Basin of Hokkaido based on the GCM's outputs and the Soil and Water Assessment Tool (SWAT) model.

6.2 Materials and Methods

6.2.1 Study Area

Hokkaido is the main northernmost island in Japan (41°21′–45°33′ N, 139°20′–148°53′ E; Fig. 6.1a). The west is the Sea of Japan, the south and the east are the Pacific Ocean, and the north is the Sea of Okhotsk. Hokkaido is famous for its relatively cool summers and cold winters. In both cases, the average temperature range in August is from 17 to 22 °C, and the average temperature in January is from −12 to −4 °C, both depending on the altitude and distance from the ocean. Unlike other major islands in Japan, Hokkaido is generally unaffected by the rainy season from June to July. The total precipitation is from 1,600 mm in the mountains of the Sea of Japan to around 800 mm around the Okhotsk Sea and inland lowlands (the lowest in Japan), while the Pacific side is about 1,100 mm.

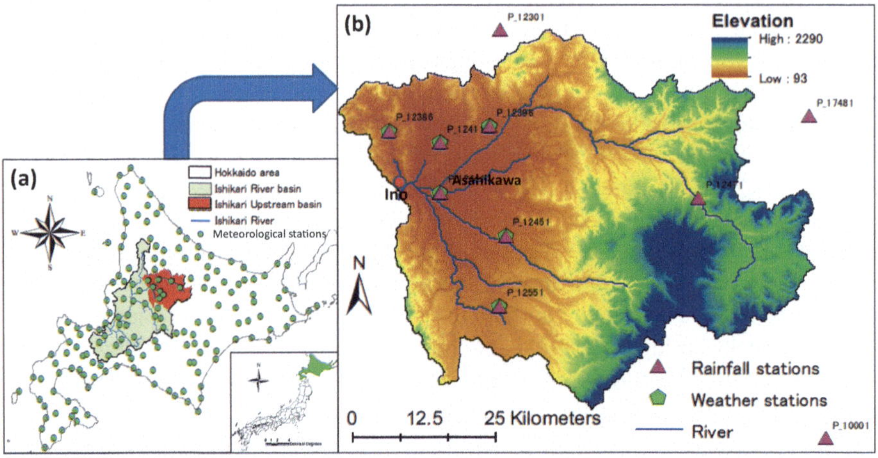

Fig. 6.1 **a** Meteorological station distributions used and **b** location of the Upper Ishikari River Basin in this study

The upper Ishikari River Basin (UIRB) is a headwater watershed of the Ishikari River and is derived from Mt. Ishikaridake (elev. 1967 m) (Fig. 6.1b). The UIRB extends from the source of the Ishikari River in the Taisetsu Mountains and to the Asahikawa city, and they mainly has Jurassic–Cretaceous rocks, serpentinite, and Cretaceous forearc sediment. The focus of this study is on the watershed area above the Ino discharge monitoring station (elev. 90.8 m), about 3,450 km^2, which accounts for about a quarter of the Ishikari River Basin. Snowmelt is the main water source of the river discharges, especially from April to May. From 1981 to 2010, the mean annual, monthly air temperatures at the Asahikawa weather station (elev. 120 m) in the warmest month (August) and the coldest month (January) were about 6.9 °C, 21.3 °C and −7.5 °C, respectively, and the mean annual precipitation is about 1042.00 mm. From the beginning of December to the end of April, the UIRB area is covered by snow for up to five months in a year.

6.2.2 Data

6.2.2.1 Precipitation Dataset

Daily precipitations at 169 stations obtained from the Automated Meteorological Data Acquisition System (AMeDAS) (Kawamoto et al. 2011) were used in this chapter. The time period was from 1980 to 2011, which is enough to reflect the latest trends. Missing values were filled in by interpolation based on neighboring days, a method that has been employed in many previous studies (Zhang et al. 2011, 2012). The Worsley likelihood ratio and the Bayesian procedures in AnClim (Stepanek 2007) were used to analyze the homogeneity for each station and results suggested that all stations selected in this study are homogeneous at >95% confidence. Four seasons were divided as: winter = December, January, February; spring = March, April, May; summer = June, July, August; autumn = September, October, November.

6.2.2.2 SWAT Model Input Datasets

Based on the ArcSWAT 2012 interface, a 50-m grid resolution (DEM) data downloaded from the Japan National and Regional Policy Bureau was used to depict the UIRB and analyze the drainage pattern of the land surface topography. Based on this DEM, the characteristics of the river network (such as the slope, length, and width of the channel) and the associated sub-basin parameters (such as the slope and slope of the terrain) were calculated. Land use is an important factor affecting runoff, evapotranspiration, and surface erosion in the basin. Soil type is one of the most important factors affecting soil moisture transport because different soil types have different soil textures and physical and chemical properties. The land-use data was developed using data from the Bureau of Land, Infrastructure, Transport, and Tourism of Japan in 2006. There were nine types of land use in UIRB (Fig. 6.2a).

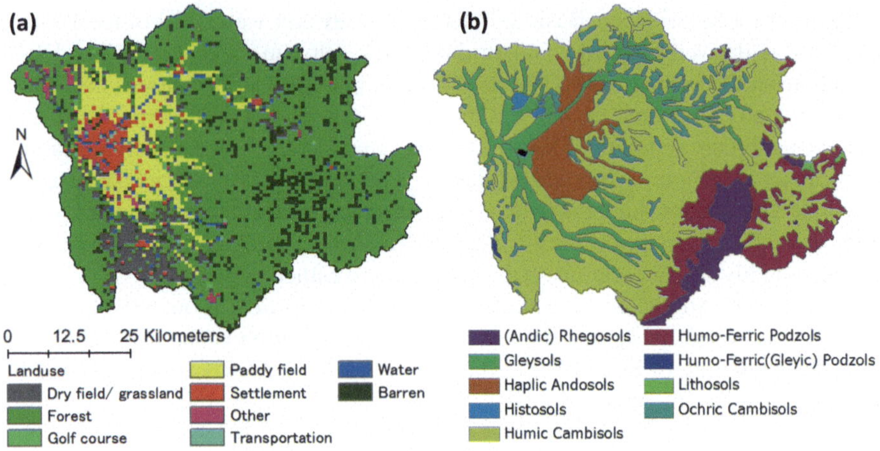

Fig. 6.2 SWAT input datasets: **a** Land use and **b** Soil

Figure 6.2b shows the soil data, which was extracted from a 1:50,000 soil map of the Fundamental Land Classification Survey developed by the Hokkaido Regional Development Bureau (www.agri.hro.or.jp/chuo/kankyou/soilmap/html/map_index.htm). Daily precipitation, maximum and minimum temperatures, wind speed, solar radiation, and relative humidity weather data were obtained from the records of rainfall and weather stations from 1981 to 2005 (Fig. 6.1b). Daily river flow data from 1995 to 2005 at Inou station data was downloaded from the website of the Ministry of Land, Infrastructure, Transport and Tourism (www1.river.go.jp) for data calibration and verification.

6.2.3 *Methods*

6.2.3.1 Statistical Analysis

The MAKESENS Microsoft Excel template developed by the Finnish Meteorological Institute (Salmi 2002) was employed to detect positive or negative trends at all 169 stations (1980–2011) for annual, seasonal, and monthly rainfall data. This software uses the Mann–Kendall trend test, a non-parametric rank-based statistical test (Kendall 1975; Mann 1945), which has been used to analyze trends in hydro-meteorological time series such as streamflow, rainfall, temperature, and water quality (Türkeş 1996; Yue et al. 2002).

To detect spatial dependence of rainfall trends among the 169 stations, we calculated Moran's I tests for spatial autocorrelation. If data exhibit spatial autocorrelation, the locations of the stations will provide information about the spatial pattern or trend of variation in these rainfall data. The global measure of spatial autocorrelation is defined as follows (Elhorst and Strijker 2003):

$$I = \frac{n \sum_{i=1}^{n} \sum_{j=1}^{n} w_{ij}(R_i - \bar{R})(R_j - \bar{R})}{(\sum_{i=1}^{n} (R_i - \bar{R})^2)(\sum_{i=1}^{n} \sum_{j=1, j \neq i}^{n} w_{ij})} \tag{6.1}$$

where R_i and R_j refer to the rainfall trend in station i and station j, respectively; $\overline{R}$ is the overall mean rainfall trend; n is the number of spatial units indexed by i and j. w_{ij} is the $(i, j)^{th}$ element of the matrix w, describing the spatial weights; that is, if station i and station j are adjacent $w_{ij} = 1$, otherwise $w_{ij} = 0$. Like the correlation, Moran's I is positive if both R_i and R_j lie on the same side of the mean (above or below); on the contrary, Moran's I is negative if one is above the mean and the other is below (O'Sullivan and Unwin 2003). GeoDa, a spatial analysis software developed by Anselin et al. (2006), was applied to calculate the degree of spatial autocorrelation among the 169 stations.

6.2.3.2 SWAT Model

The Soil and Water Assessment Tool (SWAT) model is a semi-distributed model that can be applied across the basin to simulate the quality and quantity of surface water and groundwater and to predict land use, land management practices, and environmental impacts. (Arnold et al. 1998; Narsimlu et al. 2013). In the model, hydrological response units (HRUs) were used to describe spatial heterogeneity in terms of land cover, soil type, and slope of land surface within a basin. Based on the water balance equation, the SWAT model simulates the hydrological cycle. The equation can be expressed as follows:

$$SW_t = SW_0 + \sum_{i=1}^{t} (R_{\text{day}} - Q_{\text{surf}} - E_a - w_{\text{seep}} - Q_{gw}) \tag{6.2}$$

where SW_t represents the final soil water content (mm water), SW_0 represents the initial soil water content in day i (mm water), t is the time (days), R_{day} represents the amount of precipitation in day i (mm water), Q_{surf} represents the amount of surface runoff in day i (mm water), E_a represents the amount of evapotranspiration in day i (mm water), w_{seep} represents the amount of water entering the vadose zone from the soil profile in day i (mm water), and Q_{gw} represents the amount of return flow in day i (mm water). More detailed information on SWAT model principles can be accessed from (Neitsch et al. 2005).

The model application consists of six steps: (1) data preparation; (2) watershed and sub-basin discretization; (3) HRU definition; (4) parameter sensitivity analysis; (5) calibration and verification; and (6) uncertainty analysis. In this study, the outlet of the UIRB was the Ino river discharge monitoring station.

The sensitivity analysis is performed to delineate the number of parameters that affect the fit between the simulated data and the observed data. Calibration was performed using the data from 1996–2000 and model validation was performed using the data from 2001–2005. Five years (1991–1995) were chosen as the warm-up period during which the model was allowed to initialize and then approximated to a reasonable initial value for the model state variables. The sequential uncertainty fitting algorithm (SUFI-2) was used to do the uncertainty analysis that was performed in SWAT-CUP (Abbaspour et al. 2007).

To assess the performance of model calibration, the coefficient of the determination (R^2) and Nash–Sutcliffe efficiency (NSE) between the observations and the final best simulations are calculated. The former is usually used to evaluate how accurately the model tracks the variation of the observed values. The latter measures the goodness of fit and would approach unity if the simulation is satisfactorily representing the observed data, which describes the explained variance for the observed data over time that is accounted for by the SWAT model (Green and Van Griensven 2008). R^2 ranges between 0.0 and 1.0 and higher values mean better performance. NSE indicates how well the plot of observed values versus simulated values fits the 1:1 line and ranges from $-\infty$ to 1 (Nash and Sutcliffe 1970). Larger NSE values are equivalent with better model performance. Therefore, a few standards were adopted currently for evaluating model performance. For example, Santhi et al. (2001) used the standards of $R^2 > 0.6$ and $NSE > 0.5$ to determine how well the model performed. Chung et al. (2002) used the criteria of $R^2 > 0.5$ and $NSE > 0.3$ to determine if the model result is satisfactory. In this study, $R^2 > 0.5$ and $NSE > 0.5$ were chosen as criteria for acceptable SWAT simulation.

The determination coefficient (R^2) and the Nash–Sutcliffe efficiency (NSE) between the observed and final best simulations were calculated to assess the performance of the model calibration. The NSE measures the goodness of fit would approach unity if the simulation is satisfactorily representing the observed data (Green and Van Griensven 2008), which ranges from $-\infty$ to 1 and can show how well the plot of observed values versus simulated values fits the 1:1 line (Nash and Sutcliffe 1970). Generally, $R^2 > 0.5$ and $NSE > 0.5$ were considered as criteria for acceptable SWAT water simulations. Both of them can be expressed using the following equations, respectively.

$$R^2 = \frac{\left[\sum_{i=1}^{n}\left(Q_{i,\text{obs}} - \overline{Q_{\text{obs}}}\right) * \left(Q_{i,\text{sim}} - \overline{Q_{\text{sim}}}\right)\right]^2}{\sum_{i=1}^{n}\left(Q_{i,\text{obs}} - \overline{Q_{\text{obs}}}\right)^2 * \sum_{i=1}^{n}\left(Q_{i,\text{sim}} - \overline{Q_{\text{sim}}}\right)^2} \tag{6.3}$$

$$NSE = 1 - \frac{\sum_{i=1}^{n}\left(Q_{i,\text{obs}} - Q_{i,\text{sim}}\right)^2}{\sum_{i=1}^{n}\left(Q_{i,\text{obs}} - \overline{Q_{\text{obs}}}\right)^2} \tag{6.4}$$

where $Q_{i,\text{obs}}$ is the observed data value at time unit n, $Q_{i,\text{sim}}$ is the simulated data value at time unit n, and $n = 1, 2, 3, \ldots\ldots, t$, $.\overline{Q_{\text{obs}}}$. is the mean observed data for the entire evaluation time period, $\overline{Q_{\text{sim}}}$ is the mean model simulated data for the entire evaluation time period. R^2 ranges from 0.0 to 1.0 and higher values mean better performance

6.2.3.3 GCM Data and NCEP Predicators

GCM is the most advanced tool currently available to simulate the response of the global climate system to increased greenhouse gas concentrations, which can provide global climate variables for different emission scenarios. Since some studies (He et al. 2011; Tatsumi et al. 2014) indicated that the HadCM3 (Hadley Centre Coupled Model, version 3) GCM was selected as a representative of the Japanese region, the output of HadCM3 GCM was considered suitable for this study.

Large-scale predictor variables' information, including the National Centers for Environmental Prediction (NCEP_1961-2001) reanalysis data for the calibration and validation and HadCM3 (Hadley Centre Coupled Model, version 3) GCM (H3A2a_1961-2099 and H3B2a_1961-2099) data for the baseline and climate scenario periods, were downloaded from the Canadian Climate Change Scenarios Network (http://www.cccsn.ec.gc.ca/). The NCEP reanalysis predictor contains 41 years of daily observed predictor data, derived from the NCEP reanalyzes, normalized over the complete 1961–1990 period. These data were interpolated to the same grid as HadCM3 (2.5 latitude × 3.75 longitude) before the normalization was implemented. The HadCM3 GCM predictor contains 139 years of daily GCM predictor data, derived from the HadCM3 A2 (a) and B2 (a) experiments, normalized over the 1961–1990 period. The predictors of the NCEP and HadCM3 GCM experiments with descriptions are presented in Table 6.1.

From the Canadian Climate Change Scenarios Network (http://www.cccsn.ec.gc.ca/), we downloaded the National Centers for Environmental Prediction (NCEP_1961-2001) reanalysis data and the HadCM3 (Hadley Centre Coupled Model, version 3) GCM (H3A2a_1961-2099 and H3B2a_1961-2099) data. The former was used for calibration and validation, and the latter was used for the baseline and climate scenario periods. The NCEP reanalysis predictor contains 41 years of daily observed predictor data that was reanalyzed by NCEP and normalized throughout the period 1961–1990. The daily GCM predictor data of 139 years were used as the HadCM3 GCM predictor, which were derived from the HadCM3 A2 (a) and B2 (a) experiments. Table 6.1 lists the predicted indicators and their descriptions for the NCEP and HadCM3 GCM experiments.

6.2.3.4 Downscaling Techniques

Since the resolution of GCM output data is too rough to be directly applied to the impact assessment for a specific area, it is necessary to downscale the GCM output data to bridge the gap between spatial and temporal resolution. In general, there are two downscaling techniques. One is statistical downscaling, in which statistical relationships are established by observed datasets statistical relationships between large variables (such as atmospheric surface pressure) and local variables (such as wind speed at a particular location). This relationship is then used for GCM data to obtain local variables from the GCM output. The other is dynamic downscaling, which simulates local conditions in more detail because the GCM output is used to

Table 6.1 Daily predictor variable held in the grid box data archive

Variable	Description
temp	Mean temperature at 2 m
mslp	Mean sea level pressure
p500	500 hPa geopotential height
p850	850 hPa geopotential height
rhum	Near surface relative humidity
r500	Reative humidity at 500 hPa height
r850	Relative humidity at 850 hPa height
shum	Near surface specific humidity
s500	Specific humidity at 500 hPa height
s850	Specific humidity at 850 hPa height
Derived variable	The following variables have been derived using the geostrophic approximation
[a]_f	Geostrophic air flow velocity
[a]_z	Vorticity
[a]_u	Zonal velocity component
[a]_v	Meridional velocity component
[a]zh	Divergence
[a]th	Wind direction

Notice The derived variables have been derived using the geostrophic approximation
[a]Refers to different atmospheric levels: the surface (p_), 850 hPa height (p8) and 500 hPa height (p5)

drive regional numerical models of higher spatial resolution. In this study, the statistical downscaling method was used because it is simpler and less computationally intensive than the dynamic downscaling method (Wilby et al. 2000).

The statistical downscaling contains many methods such as regression methods, weather pattern-based approaches, and stochastic weather generators. The Statistical DownScaling Model (SDSM), which is a hybrid of a stochastic weather generator and a multivariate regression method for generating local meteorological variables at a location of interest (Wilby et al. 2002), was applied to assess the impacts of climate change under future climate scenarios in this study. Based on a combination of multi-linear regressions and a weather generator, the SDSM simulates daily climate data for current and future time periods by calculating the statistical relationships between predictand and predictor data series. As shown in Fig. 6.3, the procedures of the SDSM downscaling mainly contain six steps. The quality control was used to identify gross data errors, specify missing data codes, and outliers prior to model calibration. The main purpose of the screen variable option is to choose the appropriate downscaling predictor variables. The calibrate model operation constructs downscaling models based on multiple regression equations, given daily weather data (the predictand) and regional–scale, atmospheric (predictor) variables. In this study, the

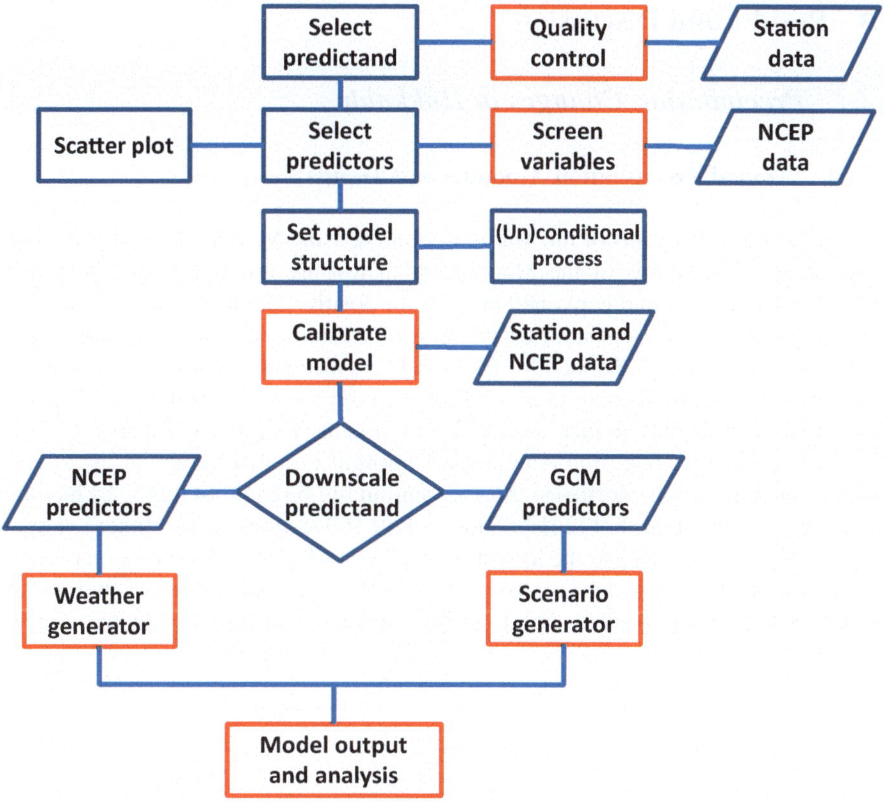

Fig. 6.3 SDSM downscaling procedures (modified from Wilby and Dawson 2007)

ordinary least-squares optimization was selected to evaluate the SDSM optimizes. The calibrated model was used to generate synthetic daily weather series using the observed (or NCEP reanalysis) atmospheric predictor variables and regression model weights. Then, the generated weather series was compared with observed station data to validate the model.

The Statistical Downscaling Model (SDSM) is a mixture of stochastic weather generators and multiple regression methods for generating local meteorological variables at locations of interest (Wilby et al. 2002), which was employed to assess the impacts of climate change in UIRB. Figure 6.3 shows the process of SDSM downscaling, which mainly consists of six steps including quality control, screen variables, calibration model operation, weather generator, scenario generator, and model output and analysis.

6.3 Results and Discussion

6.3.1 *Precipitation Changes in Hokkaido*

6.3.1.1 Annual Precipitation Amounts and Trends

Table 6.2 shows the results of the Mann–Kendall test and Moran's *I* for 169 stations in Hokkaido. The trends of the Mann–Kendall test are reported at the 90% and 95% confidence level and explained in terms of positive and negative implications. Figures 6.4 and 6.5 show spatial distribution of seasonal and monthly rainfall and trends for 169 stations in Hokkaido, 1980–2011, respectively. These changes are in line with the result of Kimoto et al. (2005) and Fujibe et al. (2005), both of whom showed mean precipitation increases over long time periods across Japan.

As shown in Fig. 6.4a, the annual mean rainfall is much higher in the southwest as compared to the northeast. Morino station (42.632° N, 141.245° E) had the highest mean annual rainfall (2213.53 mm) of all 169 stations, while Tokoro station (44.115° N, 144.037° E) had the lowest (694.75 mm). Figure 6.4b shows increasing annual rainfall at 147 stations (approximately 87% of the total number of stations) in Hokkaido. Out of 169 stations, trends are significant for 50 stations (approximately

Table 6.2 Results of Mann–Kendall test and Moran's I for 169 stations, Hokkaido

Variable	Positive	Negative	Significant 95%		Significant 90%		Moran's I
			Positive	Negative	Positive	Negative	
Annual	147	22	31	3	47	3	0.3368
Winter	103	66	17	14	26	19	0.5398
Spring	143	26	13	1	23	1	0.3884
Summer	139	30	38	0	58	0	0.3699
Autumn	116	53	0	1	5	3	0.3491
January	68	101	4	20	4	24	0.5189
February	101	68	7	9	22	14	0.5821
March	75	94	6	5	8	7	0.2505
April	69	100	0	0	0	3	0.4037
May	168	1	29	0	64	0	0.5007
June	114	55	1	0	3	0	0.4278
July	165	4	55	0	77	0	0.3506
August	89	80	0	7	1	11	0.4181
September	138	31	2	2	8	2	0.2201
October	53	116	0	3	1	9	0.4065
November	115	54	3	3	6	6	0.5721
December	122	47	9	5	28	10	0.5316

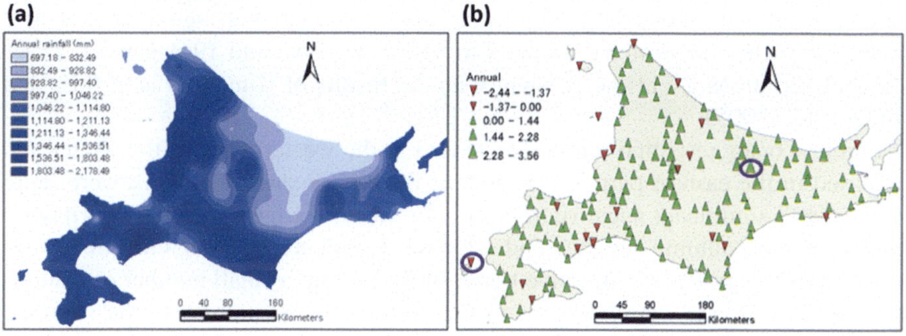

Fig. 6.4 Annual rainfall amounts (**a**) and trends (**b**) for 169 stations in Hokkaido from 1980 to 2011

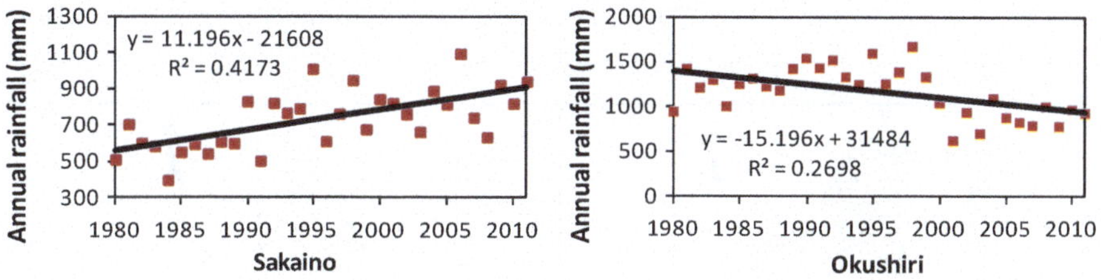

Fig. 6.5 Annual changes of rainfall amounts at Sakaino and Okushiri station from 1980 to 2011

30%, 47 positive and 3 negative) at the 90% confidence level and for 34 stations (approximately 20%, 31 positive and 3 negative) at the 95% confidence level (see Table 6.2). These stations are distributed widely across Hokkaido (see Fig. 6.4b), suggesting annual rainfall increased overall from 1980 to 2011. Sakaino station (43.707° N, 143.643° E; Z = 3.55) and Okushiri station (42.248° N, 139.557° E; Z = −2.45) represent the largest positive and negative trends, respectively, whose annual rainfall trends from 1980 to 2011 are shown in Fig. 6.5. Sakaino station, located in an area of lower average rainfall in northeast Hokkaido, saw an increase in rainfall from 500 to 900 mm from 1980 to 2011 with a peak in 2006 (approximately 1100 mm). On the other hand, Okushiri station, located in an area of higher average rainfall in the southwest of Hokkaido, saw rainfall decline from 1200 to 800 mm during 1980 to 2011.

6.3.1.2 Seasonal Precipitation Amounts and Trends

More than 60% of stations in all four seasons have an increasing trend of precipitation (see Table 6.2). The number of positive stations in spring was the highest (143, about 85%), followed by summer (139, approximately 82%), autumn (116, approximately 69%), and winter (103, approximately 61%). Among them, there were 58 stations in summer (about 34%, 58 positive and 0 negative) with significance at 90% confidence, while in winter it dropped to 45 (26 positive and 19 negative). These seasonal variations are also consistent with the results of Kimoto et al. (2005) and Fujibe et al. (2005).

Figure 6.6 suggests that rainfall decreased in the western part of the winter and increased in the eastern part. The results of the winter show that there were more stations with significant negative trends on the western edge. A similar distribution was found in the autumn, but the trend value was less than the winter. Only three sites in the eastern region had a negative trend. In the spring, a small number of stations with negative trends were scattered in the northwest. In contrast, in the summer, some stations with negative trends tended to be scattered in the southeast, especially on the southeastern edge. In addition, stations with obvious summer trends were mainly found in the central region. Figure 6.6 also clearly illustrates the seasonal differences in precipitation distribution. Precipitation was mainly concentrated in summer (ranging from 233 to 751 mm per year) and autumn (ranging from 218 to

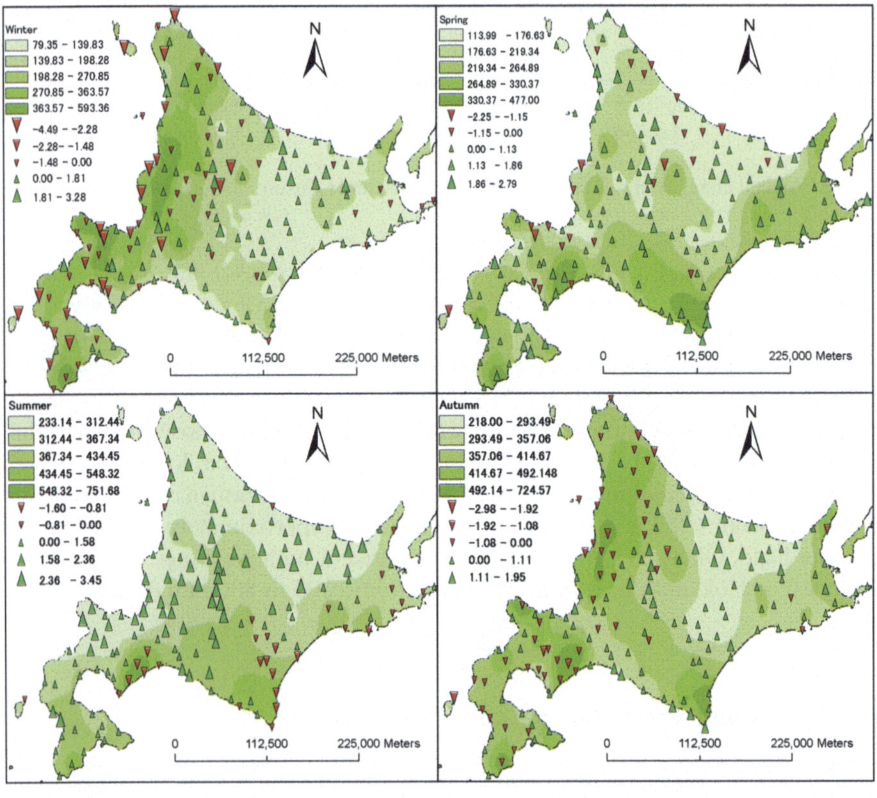

Fig. 6.6 Spatial distribution of rainfall amounts (mm) and trends (the value of Z) at seasonal scale in Hokkaido (1980–2011)

724 mm per year), while precipitation decreased in winter and spring. In winter and autumn, precipitation in the northwest was higher than in the southeast, indicating that most stations with negative trends were located in areas with high precipitation, while stations with positive trends tended to cluster in areas with lower precipitation.

6.3.1.3 Monthly Precipitation Amounts and Trends

On the monthly time scale, the increase in precipitation occurred mainly in more than 100 sites in February, May, June, July, September, November, and December (see Table 6.1). Among them, May and July had the almost all stations with a positive trend, and there were 77 stations (approximately 46%, 77 positive and 0 negative) at the 90% confidence level in July. Three months including January, April, and October had more than 100 stations with a negative trend. Among them, the number of stations with a negative trend in October was the largest (116 stations), accounting for 69%. Yamada et al. (2012) also reported an increase in precipitation from June to August and Nomoto (2003) also reported a decrease in precipitation in January.

Figure 6.7 shows that there were significant spatial differences in precipitation trends in different months. The spatial distribution of precipitation trends indicates that precipitation in most regions had decreased in January, March, April, and October, but had increased in other months, especially in May and July. Among them, the stations with a negative trend in Hokkaido in January and March were evenly distributed, mainly observed in the northwestern regions of February, April, October, November, and December, while in the southeast in June and August. Positive trends dominated in September, with the exception of the southwest corner of Hokkaido. Figure 6.7 also indicates that precipitation ranged from 31 to 162 mm in January, from 16 to 154 mm in February, from 21 to 136 mm in March and from 35 to 179 mm in December. In January, February, October, November, and December, there was more precipitation in the northwestern part of Hokkaido, similar to autumn and winter. In addition, like spring and summer, the winter and autumn rainfall in the southeastern region in April, May, June, July, August, and September was higher than in the northwest. In areas with high rainfall, most stations with negative trends were observed.

Table 6.2 also shows Moran's *I*, the measure of spatial relationship, for annual, seasonal, and monthly rainfall. Results show that monthly rainfall had the highest autocorrelation of 0.58 in February and the lowest autocorrelation of 0.22 in September. Seasonal precipitation shows the highest autocorrelation during winter and the lowest autocorrelation during autumn. Low correlation during autumn is likely caused by the high variability of the linear trend. Summer precipitation in Hokkaido is largely affected by local convection and orographic effects. Accordingly, in complex mountainous terrain, adjacent weather stations could exhibit different precipitation amounts and levels of intensity in the summer season.

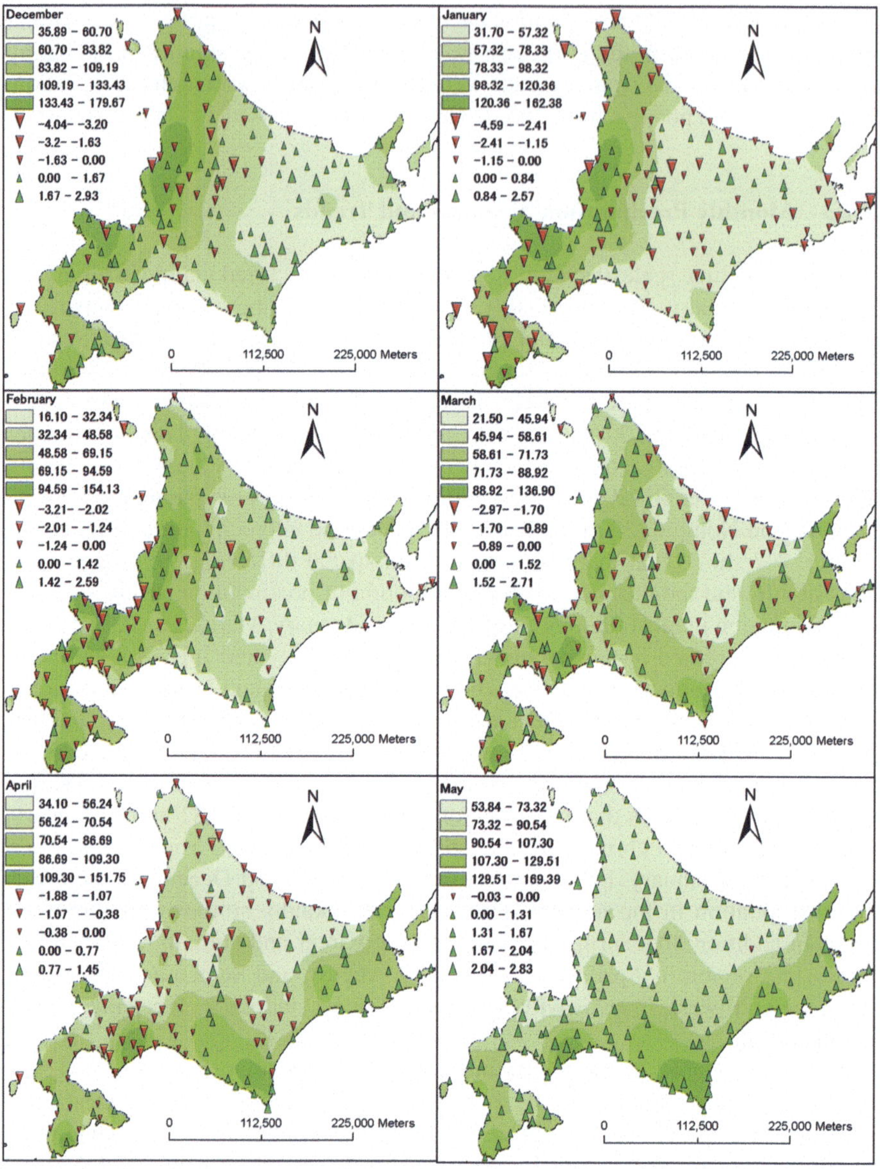

Fig. 6.7 Spatial distribution of rainfall amounts (mm) and trends (the value of Z) at monthly scale in Hokkaido (1980–2011)

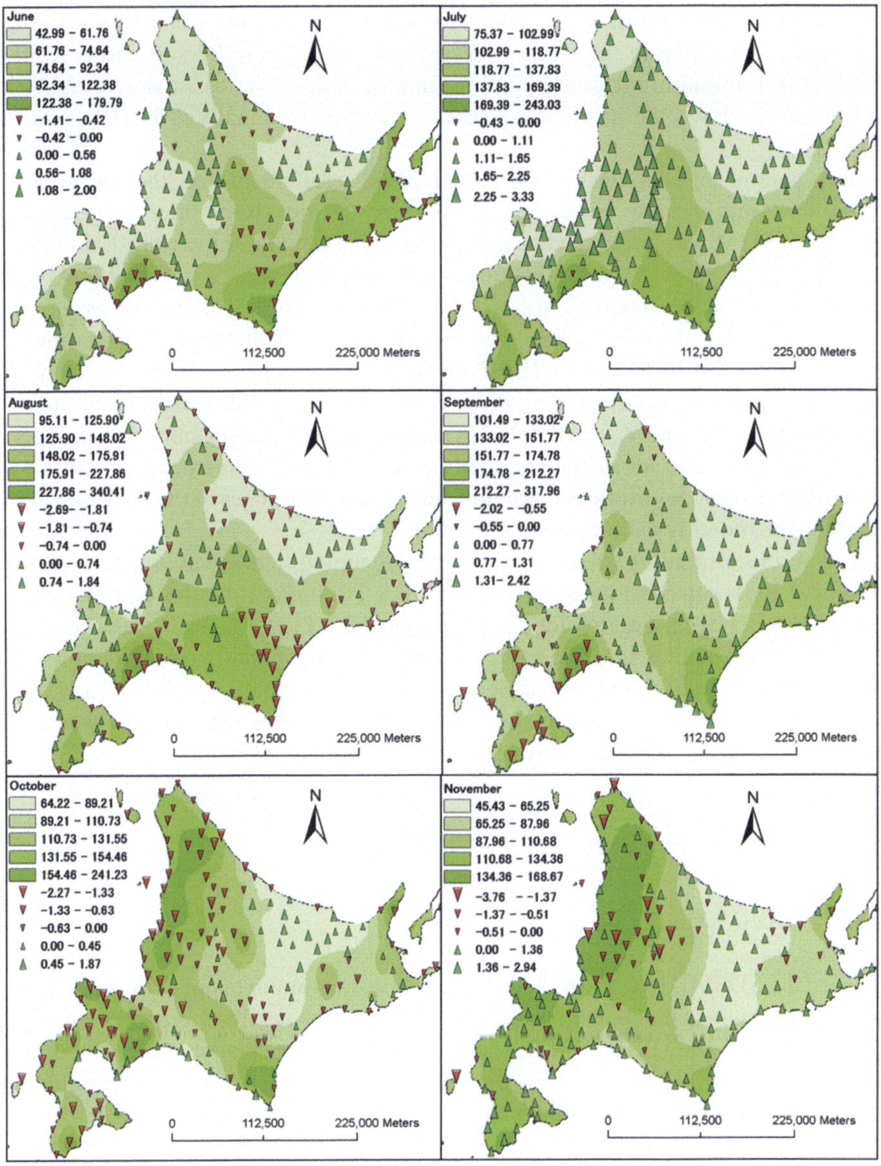

Fig. 6.7 (continued)

6.3.2 SWAT Calibration and Validation

In the discretization procedure, each available water flow gauged station was taken as a sub-basin outlet, and a threshold area of 10,000 ha (minimum area drained through a cell for the latter to be defined as a stream cell) was selected to discretize the watershed into uniform sizes. The Inou station was the only water flow gauging station used to calibrate the final SWAT model. There were 22 sub-basins in the UIRB. The minimum, maximum, and mean elevation of the basin were 91, 2290, and 608.2 m above the average sea level, respectively. Finally, 100 HRUs were created.

The parameters responsible for streamflow assessment for the catchment, viz. r_CN2.mgt (curve number), v__ALPHA_BF.gw (base flow alfa factor), v__GW_DELAY.gw (groundwater delay time), v__GWQMN.gw (threshold depth of water in shallow aquifer required for return flow), v__GW_REVAP.gw (groundwater "revap" coefficient), r__SOL_AWC.sol (soil available water capacity), r__SOL_K.sol (soil hydraulic conductivity) v__ESCO.hru (soil evaporation compensation factor), v__CH_N2.rte (manning roughness for main channel), V__SFTMP.bsn (Snowfall temperature), V__SMFMX.bsn (Maximum melt rate for snow during years), and V__SMFMN.bsn (Minimum melt rate for snow during years) have been considered for model parameterization and calibration process. Table 6.3 shows the ranking, description, minimum, and maximum ranges of the 12 parameters fitted for the monthly calibration in the SUFI-2.

The observed monthly streamflow at the Inou gauging station was used to calibrate the model, the result of which is shown in Fig. 6.8. For both calibration and validation periods, the slope of the regression line is close to the 1:1 line (the black dashed line in Fig. 6.8), which suggests a good agreement between the monthly observed flow and the simulated flow. The NSEs during calibration and verification were 0.87 and 0.86 (see Fig. 6.8), respectively, which also indicates a reasonable agreement between the calibration results and the monthly observed flow and simulated flow.

6.3.3 Climate Projects

6.3.3.1 SDSM Validation

We used the SDSM 4.2 to downscale precipitation, maximum temperature, and minimum temperature at the Asahikawa station, which can be regarded as a representative of all stations in the UIRB because the basin is relatively small compared with the GCM's resolution. Figures 6.9 shows the simulated daily maximum temperature, minimum temperature, and precipitation performance during calibration and verification, respectively, indicating a good agreement between simulated and observed daily maximum and minimum temperature values, but very bad in daily precipitation. It may be because rainfall predictions have greater uncertainty than temperature predictions (Bader et al. 2008; Wilby and Dawson 2007). However, as shown in Fig. 6.10, a relatively good agreement between simulated and observed monthly precipitation was found in both calibration and verification periods.

Figure 6.11 shows the comparison between observed average long term mean monthly precipitation, and maximum and minimum temperature with corresponding simulations, which suggests that the SDSM model can generally replicate the basic pattern of observations, especially for temperature. Based on the HadCM GCM

Table 6.3 Parameter global sensitivity ranking and final auto-calibration results

Rank	Parameter	Description	Optimal value	Lower bound	Upper bound
1	SFTMP	Snowfall temperature (°C)	4.358	−5	5
2	ESCO	Soil evaporation compensation factor	0.307	0	1
3	GW_REVAP	Groundwater "revap" coefficient	0.694	0	1
4	SOL_K	Saturated hydraulic conductivity	−0.666	−0.8	0.8
5	SOL_AWC	Available water capacity of the soil layer (mm H_2O mm^{-1} soil)	0.184	−0.5	1
6	GWQMN	Shallow aquifer required for the return flow to occur (mm)	0.112	0	2
7	GW_DELAY	Groundwater delay (days)	68.25	0	1000
8	ALPHA_BF	Baseflow alpha factor (days)	0.059	0	1
9	CH_N2	Manning's "n" value for the tributary channels	0.164	0	0.3
10	SMFMN	Minimum melt rate for snow during years (mm $°C^{-1}d^{-1}$)	4.03	0	8
11	SMFMX	Maximum melt rate for snow during years (mm $°C^{-1}d^{-1}$)	1.102	0	8
12	CN2	Initial SCS runoff curve number for moisture condition II	−0.163	−0.2	0.8

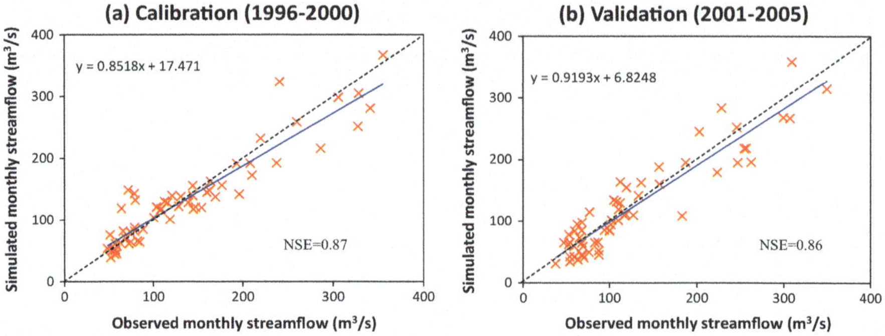

Fig. 6.8 Scatter-plot of simulated versus observed monthly streamflow at the Inou gauging station during **a** calibration and **b** validation. The dashed lines are the 1:1 lines

predictor variables for the two Emissions Scenarios (SRES, A2a and B2a), we downscaled the climate scenario for the future periods in the UIRB. Total 20 ensembles were used to calculate the future three 20-year periods, including the 2030s (2020–2039), 2060s (2050–2069), and 2090s (2080–2099). A2a describes a highly heterogeneous future world with a regionally oriented economy, whose main drivers are high population growth rates, increased energy use, land-use change, and slow technological change. B2a is also regional, but it is moving toward environmental protection and social equity.

6.3.3.2 Future Temperature

Figure 6.12 shows the changes in monthly, seasonal and annual mean maximum temperature for A2a scenario, B2a scenario, and (A2a-B2a) scenario during the future periods 2030s, 2060s, and 2090s based on the baseline period (1981–2000) at the Inou station, suggesting an increasing trend for both scenarios (A2a and B2a) in 2030s, 2060s, and 2090s, and increases during 2090s in the A2a scenario are much bigger than the B2a scenario.

For the A2a and B2a emissions scenarios, the annual average maximum temperatures may increase by 1.80 °C and 2.01 °C, 3.41 °C and 3.12 °C, and 5.69 °C and 3.76 °C, respectively, during the 2030s, 2060s, and 2090s. The seasonal results show that in the 2090s, the summer has the highest increase in the A2a and B2a emissions scenarios, reaching 6.27 °C and 3.96 °C, respectively, and the lowest in autumn, which is about 5.24 °C and 3.59 °C, respectively. In the 2090s, the A2a (about 7.40 °C) and B2a (about 4.59 °C) emissions scenarios showed the largest increase in the monthly average maximum temperature in August. Figure 6.9 also illustrates that increases in maximum temperature during 2090s in the A2a scenario are at least one degree higher than that in the B2a scenario for all months, but increases in maximum temperature during 2030s in the A2a scenario are lower than that in the B2a scenario for most months, especially in April (up to −1.59 °C).

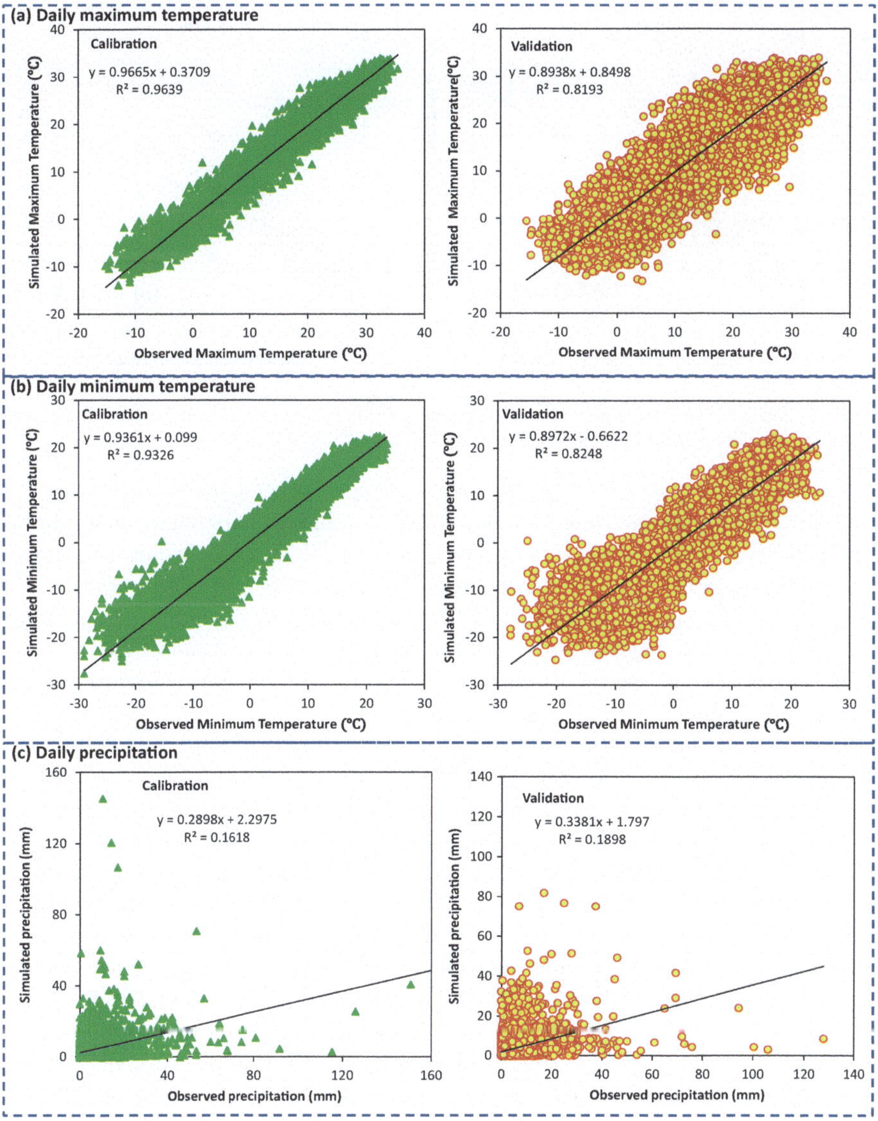

Fig. 6.9 Scatter-plot of simulated versus observed **a** daily maximum temperature, **b** daily minimum temperature, and **c** daily precipitation during calibration and validation periods

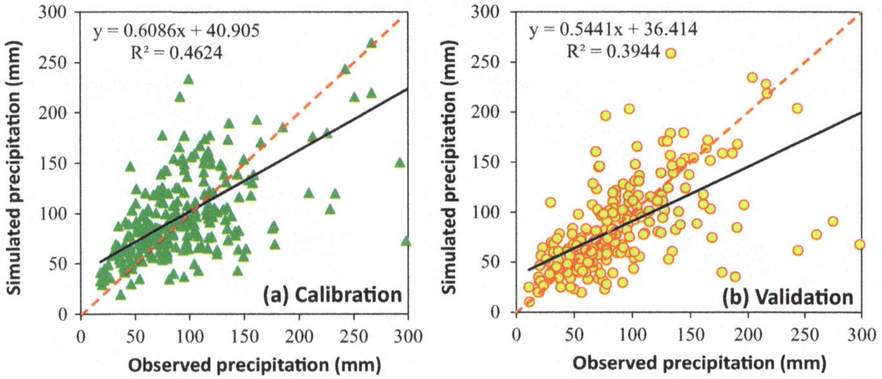

Fig. 6.10 Scatter-plot of simulated versus observed monthly precipitation during calibration (**a**) and validation (**b**). The dashed lines are the 1:1 lines

Figure 6.13 shows the changes in monthly, seasonal, and annual mean minimum temperature for A2a scenario, B2a scenario, and (A2a-B2a) scenario during the future periods 2030s, 2060s, and 2090s based on the baseline period (1981–2000) at the Inou station, the patterns of which are like the changes of maximum temperature. Generally, an increasing trend was found for both scenarios (A2a and B2a) in 2030s, 2060s, and 2090s, and increases during 2090s in the A2a scenario are much bigger than the B2a scenario.

The average annual minimum temperatures of 2030s, 2060s, and 2090s may increase by 1.41 °C and 1.49 °C, 2.60 °C and 2.34 °C, and 4.20 °C and 2.93, respectively, under the A2a and B2a emission scenarios. In terms of seasonality, winter has the largest increase in each case, followed by summer, autumn, and spring. In the 2090s, for the A2a and B2a scenarios, the average minimum temperature in winter may increase by 5.17 °C and 2.66 °C, respectively. As in the case of the monthly simulation, the minimum temperature for all twelve months of both scenarios tends to increase during all future time periods. In the 2090s, the A2a scenario had the largest increase in August (about 6.74 °C), followed by January (about 6.05 °C) and July (about 5.73 °C). In the B2a scenario, January has the largest increase (around 4.40 °C) in the 2090s, followed by August (about 4.24 °C) and July (about 3.82 °C). Figure 6.10 also illustrates that increases in minimum temperature during 2090s in the A2a scenario are at least half degree higher than that in the B2a scenario for all months, but increases in minimum temperature during 2030s in the A2a scenario are lower than that in the B2a scenario for most months, especially in March (up to −0.51 °C).

6.3.3.3 Future Precipitation

Figure 6.14 shows the changes in monthly, seasonal and annual mean precipitation for A2a scenario, B2a scenario and (A2a-B2a) scenario during the future periods 2030s, 2060s, and 2090s based on the baseline period (1981–2000) at the Inou station, which indicates that the average annual precipitation of A2a and B2a emissions scenarios

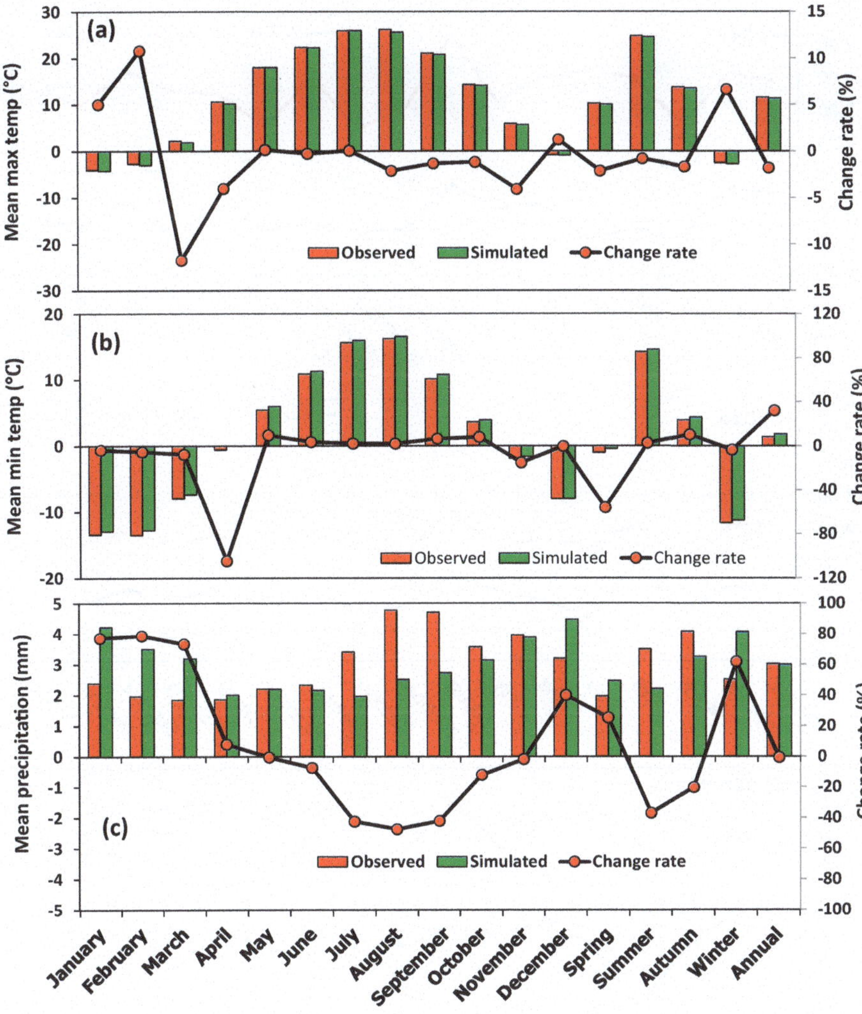

Fig. 6.11 Comparison between observed and generated mean daily precipitation and maximum and minimum temperature in the time step for the Inou station. **a** maximum temperature (°C), **b** minimum temperature (°C), and **c** precipitation (mm)

for the next 2030s, 2060s, and 2090s may decrease by 5.78% and 8.08%, 10.18% and 12.89%, 17.92%, and 11.23%, respectively, revealing a significant downward trend in future precipitation in the UIRB area. On a seasonal scale, except for the A2a scenario in the winter of the 2030s and 2060s and the spring of the 2030s, the average precipitation in all seasons may decline for both scenarios. Among them, the decline in autumn is the largest, the value of which is up to 9.52%, 20.12%, and 25.22 for the 2030s, 2060s, and 2090s, respectively, for A2a scenario, and 12.49%, 14.66%, and 20.49% for B2a scenario, followed by summer.

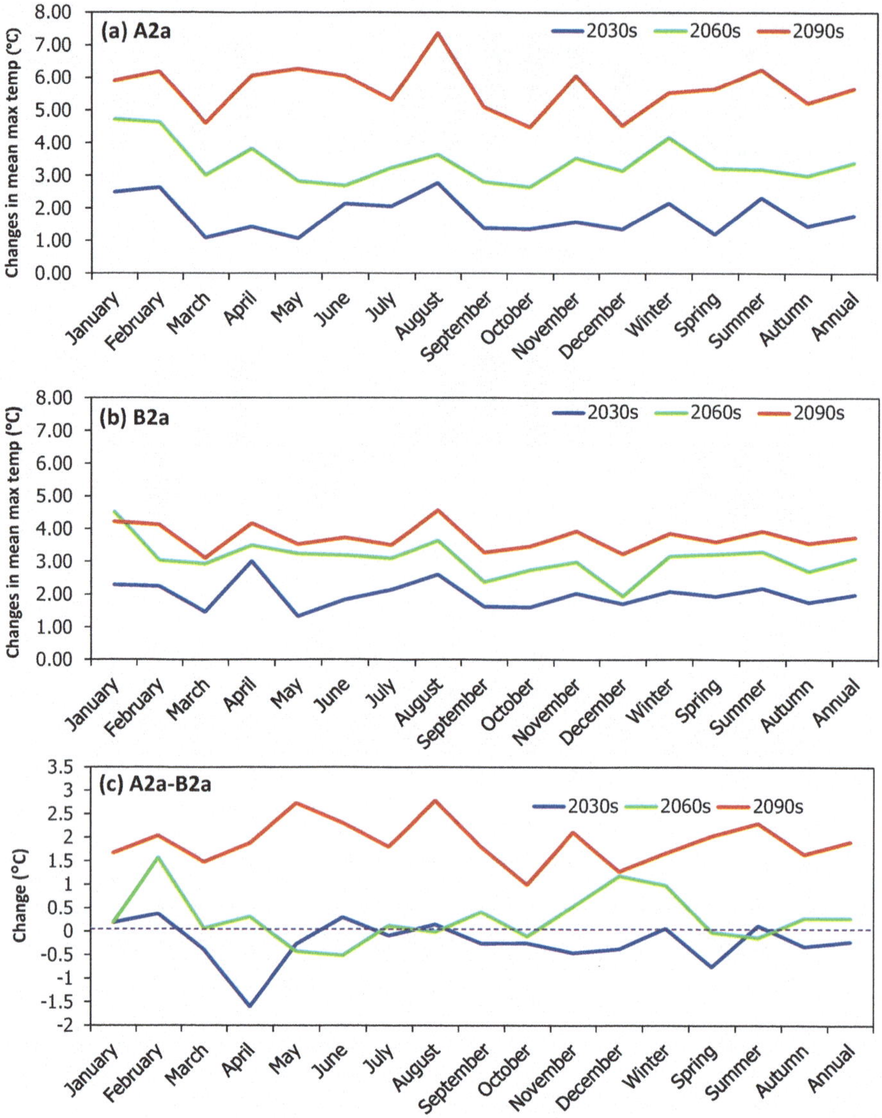

Fig. 6.12 Changes in monthly, seasonal, and annual mean maximum temperature for **a** A2a scenario, **b** B2a scenario, and **c** A2a-B2a scenario during the future periods 2030s, 2060s, and 2090s based on the baseline period (1981–2000) at the Inou station

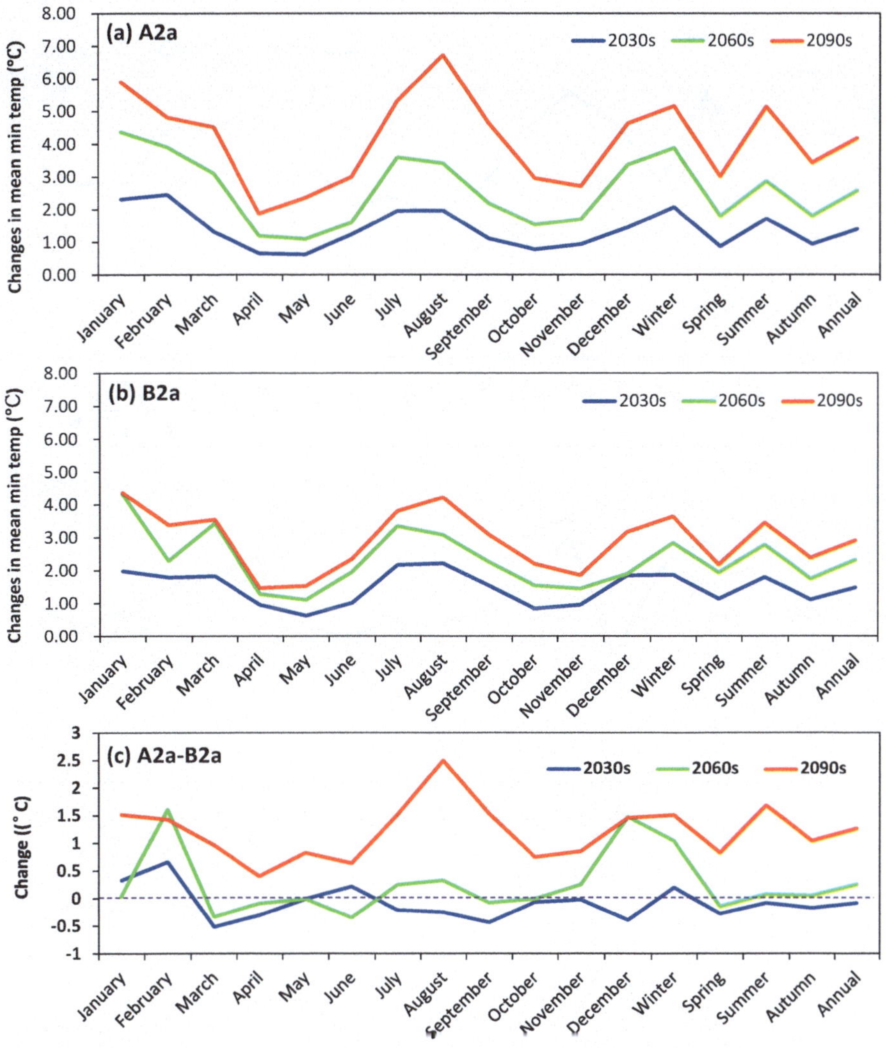

Fig. 6.13 Changes in monthly, seasonal, and annual mean minimum temperature for **a** A2a scenario, **b** B2a scenario, and **c** A2a-B2a scenario during the future periods 2030s, 2060s, and 2090s based on the baseline period (1981–2000) at the Inou station

Figure 6.14 also shows that there is a mixed trend for the future monthly precipitation for both scenarios. In the 2030s, except for February, April, May, and December under A2a scenario, and February and May under B2a scenario, the average monthly precipitation for all months may decrease. All three future periods of the A2a scenario are likely to increase in February and December compared with the baseline period. Compared with other months, a larger decline will appear in September for the 2090s for both A2a (approximately 35.47%) and B2a (approximately 27.05%)

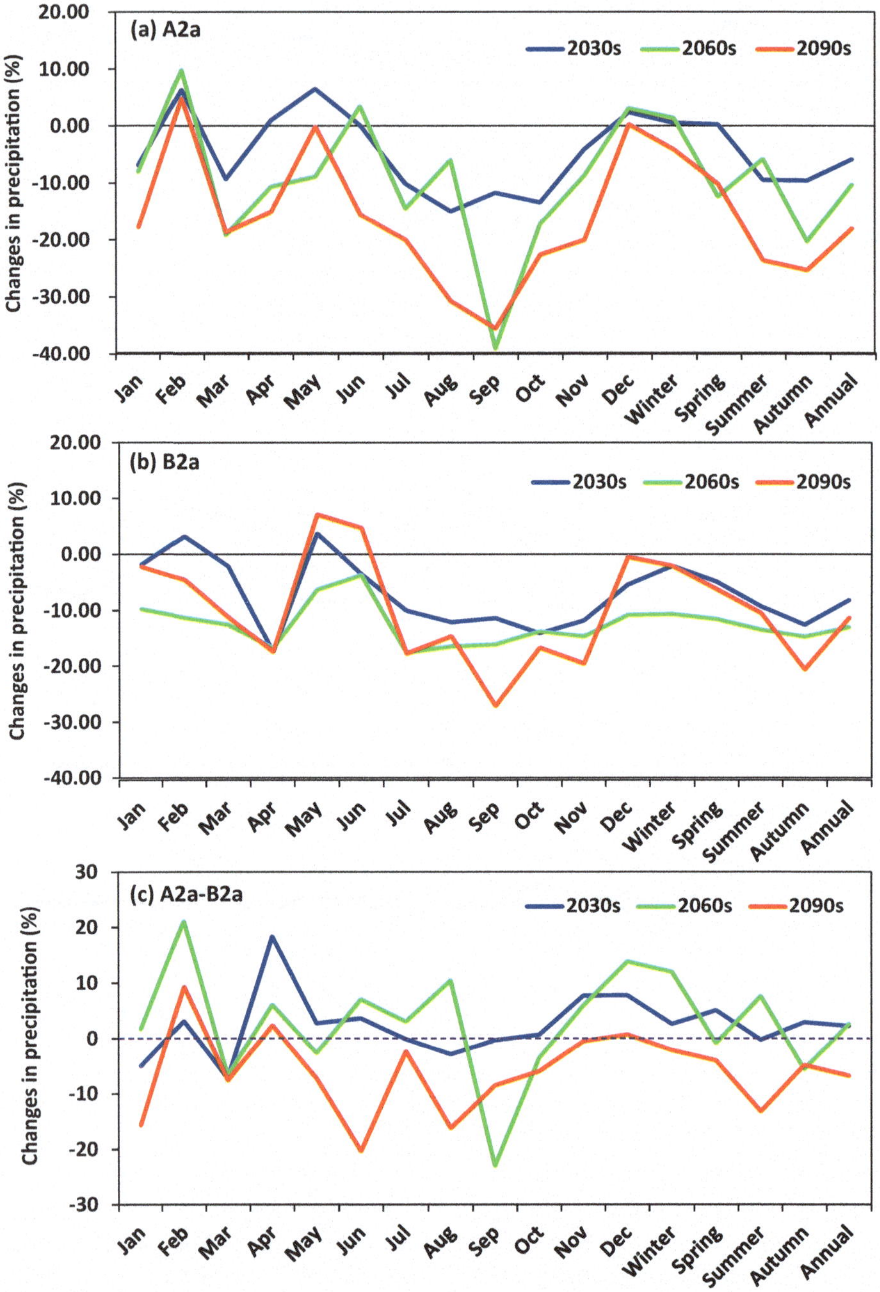

Fig. 6.14 Changes in monthly, seasonal and annual mean precipitation for **a** A2a scenario, **b** B2a scenario, and **c** A2a-B2a scenario during the future periods 2030s, 2060s, and 2090s based on the baseline period (1981–2000) at the Inou station

scenarios. Of them, September has the largest decrease for the 2060s for the A2a scenario, up to 38.90%. In addition, except for February and April, the decrease of the average precipitation for 2090s in A2a scenario is greater than the decrease in B2a scenario (see Fig. 6.14c).

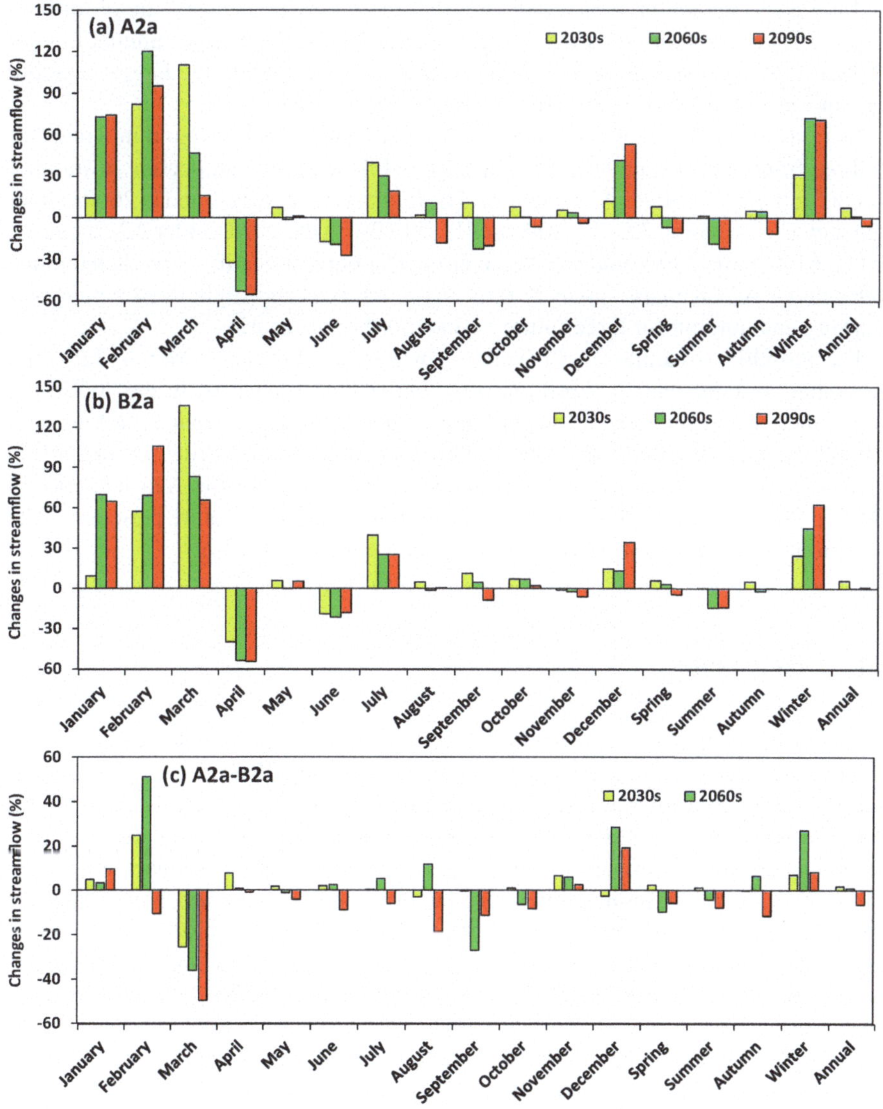

Fig. 6.15 Percentage change in mean monthly, seasonal, and annual flow volume for **a** A2a scenario, **b** B2a scenario, and **c** A2a-B2a scenario during the future periods 2030s, 2060s, and 2090s based on the baseline period (1981–2000) at the Inou gauging station

6.3.4 *Impacts on Water Flow*

Figure 6.15 shows the percentage change in mean monthly, seasonal, and annual flow volume for A2a scenario, B2a scenario and (A2a-B2a) scenario during the future periods 2030s, 2060s, and 2090s based on the baseline period (1981–2000) at the Inou gauging station, which indicates that there may be the same trend for both A2a and B2a emission scenarios. Except for the 2090s, the average annual flow of all three future periods increased in the A2a scenario. Of them, the largest increase may appear in the 2030s in the A2a scenario, up to 7.56%.

As in the case of seasonal predictions, a significant growth is found in winter for all three future periods under both scenarios, while a significant decline in summer except for the 2030s under A2a scenario. The highest increase is found in 2060s under A2a scenario, reaching 72.17%, followed by 70.89% in the 2090s under A2a scenario, and 62.61% in the 2090s under B2a scenario. The biggest decline is predicted in the 2090s under A2a scenario, up to 22.09%. Also, a mixed and slight trend will appear in spring and autumn for three future periods for both scenarios.

The monthly streamflow will increase in January, February, March, July, and December, and decrease in April and June under both scenarios. In the 2030s, the largest increase of the streamflow will appear under the B2a scenario, followed by 110.19% and 81.99% under the A2a scenario. Also, the maximum decline under B2a scenario may be as high as 40.07%, followed by 32.46% for the A2a scenario. In addition, the increase of the streamflow in March for all three future periods in A2a scenario is less than the increase in B2a scenario (see Fig. 6.15c).

6.4 Discussions

Based on the output of the HadCM3 GCM A2a and B2a climate scenarios, the temperature, precipitation, and water flow changes in the UIRB for the 2030s, 2060s, and 2090s periods were evaluated. All these data were entered into a calibrated SWAT model to calculate the water flow for all three periods in both scenarios. Several interesting questions were found. First of all, performance of the downscaling results was good in daily maximum temperature and minimum temperature, but bad in daily precipitation, which was in line with the results from Dile et al. (2013). It may be because the rainfall patterns are mainly affected by topography and GCMs cannot clearly capture these features (Bader et al. 2008; Duan et al. 2017a, b; Prudhomme et al. 2002).

Secondly, the annual precipitation in the future will decline, but there is no trend or even an upward trend in water flow. The water flow in the UIRB is expected to depend on temperature and precipitation. Remarkable decreasing precipitation will likely reduce runoff in the UIRB. However, the temperature will tend to rise, causing more snow and ice to melt, which will significantly increase the river flow. These changes are also consistent with the results of Sato et al. (2013).

Finally, although the results obtained from this study are reasonable, the uncertainty should be discussed. For example, the HaDCM3 GCM output has some uncertainty and cannot fully simulate the future, especially in precipitation (Buytaert et al. 2009). In addition, the downgrade technique also brings some uncertainty (Fowler et al. 2007; Khan et al. 2006). Thirdly, we have neglected the changes in land use and soil texture in future in the SWAT prediction, which probably cause some uncertainty in estimating water resources (Setegn et al. 2011).

6.5 Conclusions

This chapter characterized the causes of precipitation variability in Hokkaido, Japan, during 1980–2011, using Mann–Kendall test and Moran's *I*. In addition, the SWAT model was successfully applied to simulate the possible impact of climate change on water resources in the UIRB by using the GCM outputs and statistical downscaling technique. The main conclusions can be summarized as follows:

(1) Precipitation tended to increase across Hokkaido during 1980–2011 at both the annual and seasonal scales, with 147 stations (approximately 87% of the total number of stations) showing positive trends on an annual basis, followed by spring (143, approximately 85%), summer (139, approximately 82%), autumn (116, approximately 69%), and winter (103, approximately 61%).

(2) Precipitation displayed an uneven characteristic in time and space. Rainfall was mainly concentrated in summer and ranged from 233 to 751 mm and autumn with a range of 218–724 mm, while rainfall amounts were less in winter and spring. Higher rainfall occurrence in the warm seasons is consistent with observed monthly distributions. The northwest had higher rainfall than the southeast in winter, autumn and interannually, with rainfall concentrated in the southeast in both spring and summer. These results are generally in line with monthly distributions.

(3) Most stations with negative trends are located in areas with higher rainfall, such as the west in winter and autumn (see the Japan Sea side in Fig. 6.4), the southeastern fringe in summer, and scattered across the areas with lower rainfall in spring. In February, August, October, November, and December, most stations with negative trends were observed in areas with higher rainfall. This suggests that larger increases in rainfall occurred in areas with lower rainfall overall.

(4) The NSE for both the calibration and validation periods were 0.87 and 0.86, respectively, suggesting that there is reasonable agreement between the observed flow and simulated flow.

(5) Downscaling results indicate that the annual average maximum temperature may increase by 1.80 °C and 2.01 °C, 3.41 °C and 3.12 °C, and 5.69 °C and 3.76 °C, the annual average minimum temperature may be will increase by 1.41 °C and 1.49 °C, 2.60 °C and 2.34 °C, and 4.20 °C and 2.93 °C, and the annual average precipitation may be reduced by 5.78% and 8.08%, 10.18% and

12.89%, and 17.92% and 11.23% in 2030s, 2060s, and 2090s for A2a and B2a emission scenarios, respectively.

(6) In the A2a scenario, the average annual flow for all three future periods may increase, except for the 2090s. The largest increase is found in the 2030s under A2a scenario, up to 7.56%. The increase of water flow in March for all three future periods in A2a scenario is less than the increase in B2a scenario.

References

Abbaspour, K. C., Yang, J., Maximov, I., Siber, R., Bogner, K., Mieleitner, J., et al. (2007). Modelling hydrology and water quality in the pre-alpine/alpine Thur watershed using SWAT. *Journal of Hydrology, 333,* 413–430.

Anselin, L., Syabri, I., & Kho, Y. (2006). GeoDa: An introduction to spatial data analysis. *Geographical Analysis, 38,* 5–22.

Arnold, J. G., Srinivasan, R., Muttiah, R. S., & Williams, J. R. (1998). Large area hydrologic modeling and assessment part I: Model development1. *JAWRA Journal of the American Water Resources Association, 34,* 73–89.

Bader, D., Covey, C., Gutowski, W., Held, I., Kunkel, K., Miller, R., Tokmakian, R., & Zhang, M. (2008). Climate models: An assessment of strengths and limitations. A report by the US Climate Change Science Program and the Subcommittee on Global Change Research, Office of Biological and Environmental Research, Department of Energy, Washington, DC, USA, 124p.

Brown, M. E., & Funk, C. C. (2008). Food security under climate change. *Science, 319,* 580–581.

Buytaert, W., Célleri, R., & Timbe, L. (2009). Predicting climate change impacts on water resources in the tropical Andes: Effects of GCM uncertainty. *Geophysical Research Letters, 36,* L7406.

Chung, S. W., Gassman, P. W., Gu, R., & Kanwar, R. S. (2002). Evaluation of EPIC for assessing tile flow and nitrogen losses for alternative agricultural management systems. *Transactions of the ASAE, 45,* 1135–1146.

Coumou, D., & Rahmstorf, S. (2012). A decade of weather extremes. *Nature Climate Change, 2,* 491–496.

Dile, Y. T., Berndtsson, R., & Setegn, S. G. (2013). Hydrological response to climate change for Gilgel Abay River, in the lake Tana Basin-Upper Blue Nile Basin of Ethiopia. *PLoS One, 8,* e79296.

Duan, W., He, B., Sahu, N., Luo, P., Nover, D., Hu, M., et al. (2017a). Spatiotemporal variability of Hokkaido's seasonal precipitation in recent decades and connection to water vapour flux. *International Journal of Climatology, 37,* 3660–3673.

Duan, W., He, B., Takara, K., Luo, P., Nover, D., & Hu, M. (2017b). Impacts of climate change on the hydro-climatology of the upper Ishikari river basin, Japan. *Environmental Earth Sciences, 76,* 490.

Elhorst, J. P., & Strijker, D. (2003). Spatial developments of EU agriculture in the post war period: The case of wheat and tobacco. *Agricultural Economics Review, 4,* 63–72.

Fowler, H. J., Blenkinsop, S., & Tebaldi, C. (2007). Linking climate change modelling to impacts studies: Recent advances in downscaling techniques for hydrological modelling. *International Journal of Climatology, 27,* 1547–1578.

Fujibe, F., Yamazaki, N., Katsuyama, M., & Kobayashi, K. (2005). The increasing trend of intense precipitation in Japan based on four-hourly data for a hundred years. *SOLA, 1,* 41–44.

Green, C. H., & Van Griensven, A. (2008). Autocalibration in hydrologic modeling: Using SWAT2005 in small-scale watersheds. *Environmental Modelling and Software, 23,* 422–434.

He, B., Takara, K., Yamashiki, Y., Kobayashi, K., & Luo, P. (2011). Statistical analysis of present and future river water temperature in cold regions using downscaled GCMs data. *Disaster Prevention Research Institute Annuals. B, 54,* 103–111.

Hendrix, C. S., & Salehyan, I. (2012). Climate change, rainfall, and social conflict in Africa. *Journal of Peace Research, 49,* 35–50.

Iwasaki, H., & Sunaga, Y. (2009). Study of recent variation in weak rainfall over Japan using 31-year AMeDAS dataset. *Sola, 5,* 157–159.

Kawamoto, N., Oki, R., & Shimizu, S. (2011). Comparison between TRMM/PR and AMeDAS ground rain gauge network in terms of annual rainfall. In *IEEE* (pp. 2590–2593).

Kendall, M. G. (1975). Rank correlation methods: Charles Griffin. London.

Khan, M. S., Coulibaly, P., & Dibike, Y. (2006). Uncertainty analysis of statistical downscaling methods. *Journal of Hydrology, 319,* 357–382.

Kimoto, M., Yasutomi, N., Yokoyama, C., & Emori, S. (2005). Projected changes in precipitation characteristics around Japan under the global warming. *Sola, 1,* 85–88.

Konikow, L. F., & Kendy, E. (2005). Groundwater depletion: A global problem. *Hydrogeology Journal, 13,* 317–320.

Mann, H. B. (1945). Nonparametric tests against trend. *Econometrica, 13,* 245–259.

Narsimlu, B., Gosain, A. K., & Chahar, B. R. (2013). Assessment of future climate change impacts on water resources of Upper Sind River Basin, India using SWAT model. *Water Resources Management, 27,* 3647–3662.

Nash, J., & Sutcliffe, J. V. (1970). River flow forecasting through conceptual models part I—A discussion of principles. *Journal of Hydrology, 10,* 282–290.

Neitsch, S. L., Arnold, J. G., Kiniry, J. R., Williams, J. R., & King, K. W. (2005). *SWAT theoretical documentation version 2005.* Temple, TX: Blackland Research Center.

Nomoto, S. (2003). Decreases in the number of foggy days in Thailand and Japan, and possible causes. *Journal of International Economic Studies, 17,* 13–28.

Oki, T., & Kanae, S. (2006). Global hydrological cycles and world water resources. *Science, 313,* 1068–1072.

O'Sullivan, D., & Unwin, D. J. (2003). *Geographic information analysis.* Hoboken, USA: Wiley.

Stepanek, P. (2007). Software package for processing climatological data (complex solution). Retrieved October, 2019 from http://www.climahom.eu/software/docs/.

Prudhomme, C., Reynard, N., & Crooks, S. (2002). Downscaling of global climate models for flood frequency analysis: Where are we now? *Hydrological Processes, 16,* 1137–1150.

Salmi, T. (2002). *Detecting trends of annual values of atmospheric pollutants by the Mann-Kendall test and Sen's slope estimates: The Excel template application MAKESENS, Publications on air quality 31.* Helsinki: Finnish Meteorological Institute.

Santhi, C., Arnold, J. G., Williams, J. R., Dugas, W. A., Srinivasan, R., & Hauck, L. M. (2001). Validation of the SWAT model on a large river basin with point and non point source. *Journal of the American Water Resources Association, 37,* 1169–1188.

Sato, Y., Kojiri, T., Michihiro, Y., Suzuki, Y., & Nakakita, E. (2013). Assessment of climate change impacts on river discharge in Japan using the super-high-resolution MRI-AGCM. *Hydrological Processes, 27,* 3264–3279.

Setegn, S. G., Rayner, D., Melesse, A. M., Dargahi, B., & Srinivasan, R. (2011). Impact of climate change on the hydroclimatology of Lake Tana Basin, Ethiopia. *Water Resources Research, 47,* W4511.

Solomon, S. (2007). Climate change 2007: The physical science basis: Contribution of working group I to the fourth assessment report of the intergovernmental panel on climate change. Cambridge University Press.

Tatsumi, K., Oizumi, T., & Yamashiki, Y. (2014). Effects of climate change on daily minimum and maximum temperatures and cloudiness in the Shikoku region: A statistical downscaling model approach. *Theoretical and Applied Climatology, 120,* 87–98.

Türkeş, M. (1996). Spatial and temporal analysis of annual rainfall variations in Turkey. *International Journal of Climatology, 16,* 1057–1076.

Wada, Y., Van Beek, L. P., Van Kempen, C. M., Reckman, J. W., Vasak, S., & Bierkens, M. F. (2010). Global depletion of groundwater resources. *Geophysical Research Letters, 37,* L20402.
Wilby, R. L., & Dawson, C. W. (2007). SDSM 4.2-A decision support tool for the assessment of regional climate change impacts, Version 4.2 User Manual. Lancaster University: Lancaster/Environment Agency of England and Wales.
Wilby, R. L., Dawson, C. W., & Barrow, E. M. (2002). SDSM—a decision support tool for the assessment of regional climate change impacts. *Environmental Modelling and Software, 17,* 145–157.
Wilby, R. L., Hay, L. E., Gutowski, W. J., Arritt, R. W., Takle, E. S., Pan, Z., et al. (2000). Hydrological responses to dynamically and statistically downscaled climate model output. *Geophysical Research Letters, 27,* 1199–1202.
Yamada, T. J., Sasaki, J., & Matsuoka, N. (2012). Climatology of line-shaped rainbands over northern Japan in boreal summer between 1990 and 2010. *Atmospheric Science Letters, 13,* 133–138.
Yue, S., Pilon, P., & Cavadias, G. (2002). Power of the Mann-Kendall and Spearman's rho tests for detecting monotonic trends in hydrological series. *Journal of Hydrology, 259,* 254–271.
Zhang, Q., Singh, V. P., Li, J., Jiang, F., & Bai, Y. (2012). Spatio-temporal variations of precipitation extremes in Xinjiang, China. *Journal of Hydrology, 434–435,* 7–18.
Zhang, Q., Xu, C. Y., Chen, X., & Zhang, Z. (2011). Statistical behaviours of precipitation regimes in China and their links with atmospheric circulation 1960–2005. *International Journal of Climatology, 31,* 1665–1678.
Zhang, X., Zwiers, F. W., Hegerl, G. C., Lambert, F. H., Gillett, N. P., Solomon, S., et al. (2007). Detection of human influence on twentieth-century precipitation trends. *Nature, 448,* 461.

Chapter 7
Estimation of Nutrient and Suspended Sediment Loads in the Ishikari River

7.1 Introduction

The importance of surface water quality in controlling the health of aquatic ecosystems, affecting drinking water resources, and human health is increasingly recognized. Nutrients (mainly nitrogen and phosphorus) are essential for the life of animals and plants, but high concentrations cause many ecological problems (Carpenter et al. 1998; Li e 2011; Smith 1982; Sprague and Lorenz 2009). Excessive sediments reduce water quality, which has negative impacts on material fluxes, aquatic geochemistry, water quality, and channel morphology (Dedkov and Mozzherin 1992; Ishida et al. 2010; Meade et al. 1985). In order to effectively manage and protect water resources, it is important to understand nutrients and sediments dynamics (Gruber and Galloway 2008; Sprague and Lorenz 2009). Concentrations and loads of nutrients and SS are usually estimated based on models and infrequent monitoring data (Armour et al. 2009; Kulasova et al. 2012; Li et al. 2009, 2011; Ma et al. 2011). River flow data, usually obtained from sparse monitoring stations, provides the basis for estimating critical water quality components, as the concentration and load of these components typically change naturally with changes in flow. In addition, it is often difficult to obtain water-quality records representing the concentration of components in a rapidly changing water stream over a long period of time (years to decades). Therefore, the uncertainty of the composition load estimate is often high (Christensen et al. 2000; Reckhow 1994).

Japan's efforts in water management have greatly improved the estimation of water quality (Duan et al. 2013; Wang et al. 2016). For example, a national data collection network called "National Land Water Information" (http://www1.river.go.jp/) was built to offer reliable and real-time water data including river discharges, water quality, and precipitation. Since the load estimation process is complicated by the transformation bias (Ferguson 1986; Webb et al. 1997), data censoring (Gilbert 1987), and non-normality (Helsel and Hirsch 1992; Shumway et al. 2002), there is usually a bias in the estimated load of the model simulation (Johnes 2007). Meanwhile, in Japan, high suspended sediment load is increasingly recognized as an important issue

W. Duan and K. Takara, *Impacts of Climate and Human Activities on Water Resources and Quality*, https://doi.org/10.1007/978-981-13-9394-5_7

in river basin management (Mizugaki et al. 2008; Somura et al. 2012). For example, the Ishikari River Basin has long been plagued by heavy suspended sediment loads, which often result in high turbidity and riverbed erosion along rivers, including the economic and government center (e.g., Sapporo) of Hokkaido. A few studies have focused on the SS management in the Ishikari River Basin. For example, Asahi et al. (2003) argued that it is necessary to directly deal with the influence of tributaries, and the sediment discharged from the tributaries contributes to the discharge of sediment from the estuary. However, the detailed sediment sources and transportation in the Ishikari River Basin is still poorly understood.

Therefore, it is important to improve estimation methods for obtaining reliable water quality loads. Based on the Maintenance of Variance-Extension type 3 (MOVE.3) and the regression model Load Estimator (LOADEST), this chapter firstly estimated total nitrogen (TN), total phosphorus (TP), and suspended sediments (SS) loads at five sites on the Ishikari River, Japan, from January 1985 to December 2010. Then, this chapter developed a spatially explicit, regional empirical model of suspended sediments (SS, soil, and other particulate matter) for the whole Ishikari River Basin. This model is used to analyze the source and transport process of the SS. The ultimate goal of this work is to provide information and tools to help resource managers identify priority sources of pollution and reduce pollution to protect water resources and protect aquatic ecosystems in the Ishikari River Basin.

7.2 Study Area and Datasets

7.2.1 Study Area and Data Collection

Figure 7.1 shows that the Ishikari River Basin is located in the middle of Hokkaido with a total drainage area of 14,330 km^2. The Ishikari River originates from Mt. Ishikaridake (elev. 1967 m) in the Taisetsu Mountains of central Hokkaido, and flows southward into flows south into the vast Ishikari Plain and finally flows into the Sea of Japan. The length of the Ishikari River mainstream is 268 km, ranking third in Japan. At the Sapporo weather station (elev. 17 m), the monthly temperature of the warmest month (August) is about 22.0 °C, and the monthly temperature of the coldest month (January) is about −4.1 °C.

As shown in Fig. 7.1, 31 monitoring stations were selected in this chapter. Of them, 5 monitoring stations including station 7, station 10, station 17, station 20, and station 31 were used to do the trend analysis. The sites Yinou-oohashi, Yiwamizawa-oohashi, and Yishikarikakou-bashi are located in the upper, middle, and lower reaches of the mainstream, respectively, while the Akane-bashi and Umaoyi-bashi sites are located in the tributaries of Uryū and Yūbari, respectively. Typically, each site was measured and collected for water quality concentration (TN, TP, and SS) and river discharges per one month or two months from 1985 to 2010 by the National Land with Water Information (http://www1.river.go.jp/) monitoring network. A total number

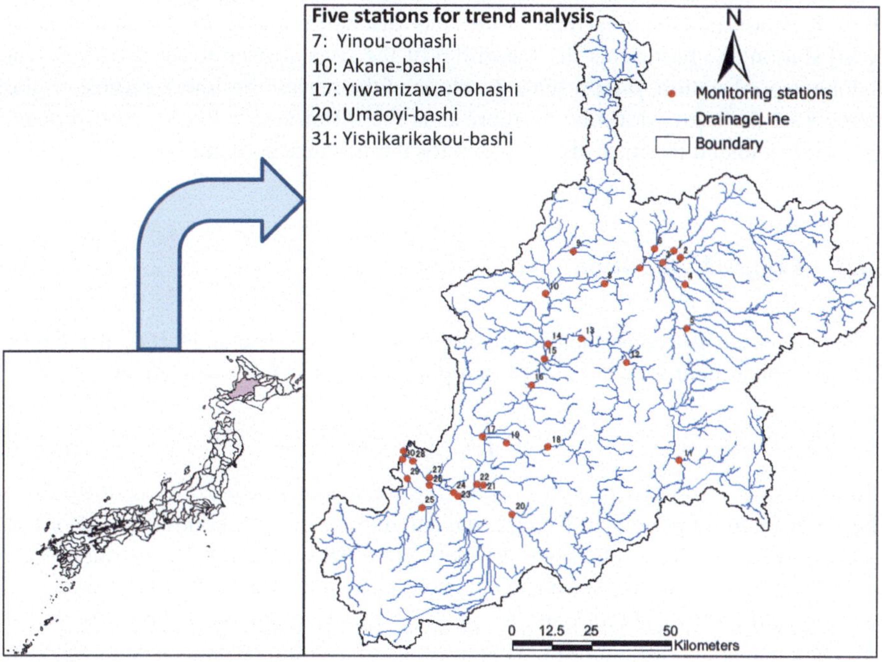

Fig. 7.1 Study area and monitoring stations for the Ishikari River

of 312 samples for each testing parameter were collected at Akane-bashi, Yinouo-ohashi, and Umaoyi-bashi stations, and 274 samples were collected for each testing parameter at Yiwamizawa-oohashi and Yishikarikakou-bashi stations. All these 31 monitoring stations were used to calibrate the SS model for the Ishikari River Basin.

7.3 Statistical Methods

7.3.1 Streamflow Extension

Based on daily streamflow values recorded at nearby and hydrologically similar index stations, a streamflow record extension method called the Maintenance of Variance-Extension type 3 (MOVE.3) (Vogel and Stedinger 1985) was applied to estimate missing flow values or to extend the record at a short-record station. The method can be expressed by the following equation:

$$Y_i = \beta X_i + b + e_i \tag{7.1}$$

where Y_i represents the logarithm of the river discharge for the ith day at the short-record station; X_i represents the logarithm of the river discharge for the ith day at the long-record station, β represents the slope of the regression-line, b represents the intercept of the regression-line, e_i represents the difference between the estimated regression-line and the measured Y value for the ith measurement.

7.3.2 Loads Estimation

In theory, based on constituent concentration (C) and discharge (Q), the SS or chemical constituent load (Ø) can be computed using the following equation:

$$\text{Ø} = \int C(t)Q(t)\mathrm{d}t \tag{7.2}$$

where t is the time period. In Eq. 7.2, the calculation needs a continuous record of concentration and discharge when estimating the SS or chemical constituent load (Ø). Although it is easy to measure river discharges at a sufficiently high frequency, it is often difficult to obtain continuous water quality data due to the expense of collecting and analyzing samples. Therefore, Eq. 7.2 could be expressed as follows:

$$L_T = \Delta t \sum_{i=1}^{n} L_i \tag{7.3}$$

where L_T represents an estimate of total water quality load, L_i represents an estimate of instantaneous water quality load, n represents the number of discrete points in time, and Δt represents the time interval represented by the instantaneous water quality load.

Then, we have calibrated the above regression model using the FORTRAN Load Estimator (LOADEST) (Runkel et al. 2004). Four statistical estimation methods including the adjusted maximum likelihood estimation (AMLE), maximum likelihood estimation (MLE), linear attribution method (LAM), and least absolute deviation (LAD) were provided in the LOADEST when carrying out the calibration procedures. Of them, when the model calibration errors (residuals) are normally distributed, both AMLE and MLE are suitable, but AMLE is the more appropriate method when containing censored data (i.e., when data are reported as less than or greater than some threshold). Other two methods including LAM and LAD are the appropriate methods when the residuals are not normally distributed. Therefore, since the input data in this chapter included censored data and the model calibration residuals were normally distributed within acceptable limits, the AMLE estimation method was chosen to calibrate at each monitoring station. The output regression model equations can be expressed as follows (Runkel et al. 2004):

$$\ln(L_i) = a + b\ln Q + c\ln Q^2 + d\sin(2\pi\, dtime) \\ + e\cos(2\pi\, dtime) + f dtime + g dtime^2 + \varepsilon \quad (7.4)$$

where L_i represents the calculated water quality load for the sample i, Q represents the stream discharge; $dtime$ represents the time, in decimal years from the beginning of the calibration period, ε represents the error, and a, b, c, d, e, f, g represent the fitted parameters in the multiple regression model of water quality loads. According to the lowest Akaike information criterion (AIC) values, some of the regression equations did not contain all terms (Sakamoto et al. 1986). The AIC could be calculated as follows:

$$AIC = 2n - 2\ln(L) \quad (7.5)$$

where n represents the number of parameters in the statistical water quality model and L represents the maximized value of the likelihood function for the estimated water quality model. The monthly and seasonal average TN, TP, and SS loads were estimated based on the above formula. Four seasons were considered as follows: winter (December, January, February), spring (March, April, May), summer (June, July, August), and autumn (September, October, November).

7.3.3 *Trend Analysis*

The Mann–Kendall test was employed to compute trends of water quality in Ishikari River Basin, which is a nonparametric rank-based statistical test (Kendall 1975; Mann 1945) and has been used to analyze trends in hydro-meteorological time series (Duan et al. 2017; Yue et al. 2002). For independent and randomly ordered on a time series $X_i\{X_i, i = 1, 2, \ldots, n\}$, the null hypothesis H_0 in Mann–Kendall test assumes that there is no trend and this is tested against the alternative hypothesis H_1, which assumes that there is a trend. The Mann–Kendall S Statistic is computed as follows:

$$S = \sum_{i-1}^{n-1} \sum_{j-i+1}^{n} \text{sign}(T_j - T_i) \quad (7.6)$$

$$\text{sign}(T_j - T_i) = \begin{cases} 1 & \text{if } T_j - T_i > 0 \\ 0 & \text{if } T_j - T_i = 0 \\ -1 & \text{if } T_j - T_i < 0 \end{cases} \quad (7.7)$$

where T_j and T_i are the water quality (TN, TP, and SS loads) variability at multiple time scales j and i, $j > i$, respectively. When $n \geq 10$, the statistic S is approximately normally distributed with the mean and variance as follows:

$$E(S) = 0 \quad (7.8)$$

The variance (σ^2) for the S-statistic is defined by

$$\sigma^2 = \frac{n(n-1)(2n+5) - \sum t_i(i)(i-1)(2i+5)}{18} \tag{7.9}$$

where t_i denotes the number of ties to extent i. The summation term in the numerator is used only if the water quality (TN, TP, and SS loads) data contains tied values. The standard test statistic Z_S is calculated as follows:

$$Z = \begin{cases} \frac{S-1}{\sigma} & \text{for } S > 0 \\ 0 & \text{for } S = 0 \\ \frac{S+1}{\sigma} & \text{for } S < 0 \end{cases} \tag{7.10}$$

The test statistic Z_S is used as a measure of significance of trend of the water quality (TN, TP, and SS loads) (e.g., Yue et al. 2002). This test statistic is used to test the null hypothesis (H_0). That is, if $|Z_S|$ is great than $Z_{\alpha/2}$, where α is the chosen significance level (e.g., 5% with $Z_{0.025} = 1.96$) then the null hypothesis is invalid implying that the trend of water quality (TN, TP, and SS loads) is significant.

7.4 Modeling Tools

7.4.1 SPARROW Model

Computer-based modeling is essential for organizing and understanding complex data related to water quality conditions and for developing management strategies and decision support tools for water resource managers (Somura et al. 2012). There are lots of hydrological and water quality models to describe and identify sources of pollutants and transportations at various spatial scales (e.g., HSPF (Jeon et al. 2011; Johanson et al. 1980), ANSWERS (Beasley et al. 1980), SWAT (Kirsch et al. 2002; Liew et al. 2012), and ANN (Rajaee et al. 2011)). Nowadays, the GIS-based watershed model SPARROW (SPAtially Referenced Regression On Watershed attributes) is very popular and widely used to evaluate the water quality in the United States (Smith et al. 1997; Alexander et al. 2000, 2007; Duan et al. 2012, 2015; McMahon et al. 2003). Figure 7.2 shows the functional linkages between the major spatial components of SPARROW models. Monitoring station flux estimates are long-term flux estimates used as response variables in the model. The flux estimates for the monitoring station are derived from a station-specific model that correlates the concentration of contaminants in each water sample with a continuous record of the water flow time series. Generally, a nonlinear least-squares multiple regression is used to describe and estimate the relationship between spatially referenced basin and channel characteristics (predictors) and in-stream water pollution loads (response) (Schwarz et al.

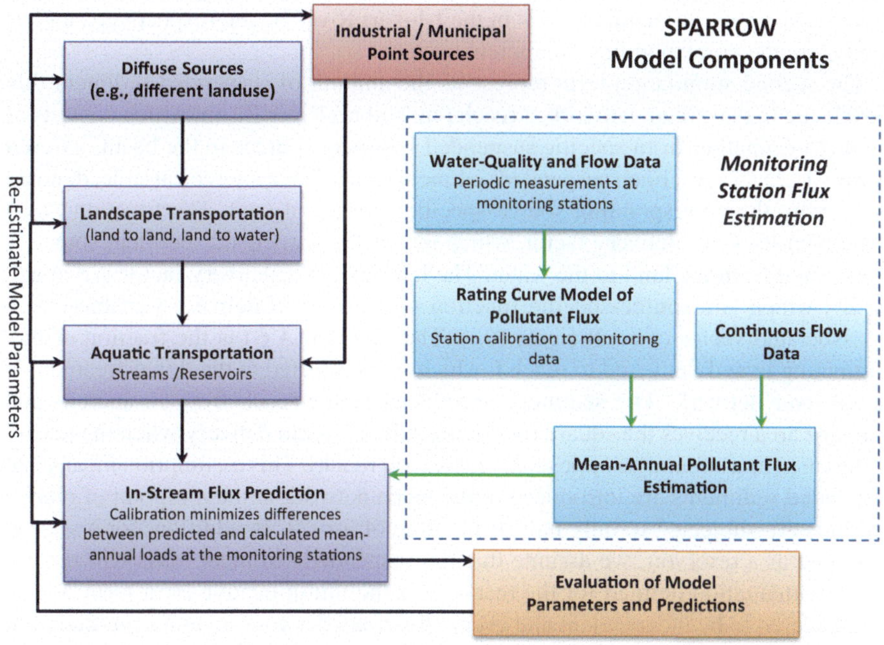

Fig. 7.2 Schematic of the major SPARROW model components (From Schwarz et al. 2006)

2006). This function can track the transportation of water quality to evaluate the pollutants' supply and attenuation in streams and reservoirs (Preston and US 2009).

The fundamental principle of SPARROW modeling can be expressed as follows (Alexander et al. 2007):

$$F_i^* = \left[\left(\sum_{j\in J(i)} F_j'\right) A\left(Z_i^S, Z_i^R; \theta_S, \theta_R\right) + \left(\sum_{n=1}^{N_S} S_{n,i}\alpha_n D_n\left(Z_i^D; \theta_D\right)\right) A'\left(Z_i^S, Z_i^R; \theta_S, \theta_R\right)\right]\varepsilon_i \tag{7.11}$$

The first summation term $\left(\left(\sum_{j\in J(i)} F_j'\right) A\left(Z_i^S Z_i^S \theta_S \theta_R\right)\right)$ is the suspended sediments flux that leaves upstream reaches and is delivered downstream to reach i, where F_j' represents the measured suspended sediment flux $\left(F_j^M\right)$ when upstream reach j is monitored and equals the given model-estimated flux (F_j^*) when it is not. $A(\cdot)$ represents the stream delivery function, which indicates suspended sediment loss processes acting on flux as it travels along the reach pathway. In other words, it defines the fraction of suspended sediment flux entering reach i at the upstream node that is delivered to the downstream node. Z^S and Z^R are functions of measured stream and reservoir characteristics in the Ishikari River Basin, respectively, and θ_S and θ_R are the corresponding coefficient vectors.

The second summation term represents the amount of suspended sediment flux introduced to the stream network at reach i in Ishikari River Basin, which consists of the flux originating from specific suspended sediment sources in the basin, indexed by $n = 1, 2, \ldots, N_S$. Each suspended sediment source has a source variable, denoted as S_n, with the corresponding source-specific coefficient (α_n). The function $D_n(\cdot)$ is the land-to-water delivery factor, which means the suspended sediment source is transported from the land to the water. The land-to-water delivery factor is denoted by Z_i^{θ}, which is a source-specific function of a vector of delivery variables, with an associated vector of coefficients (θ_D). The function $A'(\cdot)$ is the fraction of flux originating in and delivered to reach i, which is transported to the downstream node in the river network. The sediment introduced to the reach from its incremental drainage area receives the square root of the full in-stream delivery when the reach i is classified as a stream (as opposed to a reservoir reach). This assumption means that suspended sediments are introduced to the reach network at the midpoint of reach i and thus are subjected to only half of the time of arrival. In addition, for a section classified as a reservoir, we assume that the suspended sediment mass will receive the full attenuation defined for the reach. ε_i is the multiplicative error term, which is considered to be independent and evenly distributed across separate sub-basins in the intermediate drainage system between stream monitoring stations.

The suspended sediment loss in streams is modeled according to a first-order decay process. In this process, the sediment mass fraction derived from the upstream node and transported along the reach i to the downstream node is estimated as a continuous function of the mean water time of travel (T_i^S; units of time) and average water depth (D_i) in reach i. The process can be expressed as follows:

$$A\left(Z_i^S, Z_i^R; \theta_S, \theta_R\right) = \exp\left(-\theta_S \frac{T_i^S}{D_i}\right) \tag{7.12}$$

where θ_S is an estimated suspended sediment mass-transfer flux-rate coefficient in units of length time^{-1}. The rate coefficient is independent of the amount of water that is proportional to the amount of water, such as streamflow and depth (3). Based on the average water depth, the rate can be re-expressed as a reaction rate coefficient (time^{-1}).

7.4.2 *Input Data*

In this study, the input data used to construct the SPARROW model is divided into: (1) stream network data to define the watershed and catchment area in the Ishikari River Basin; (2) The suspended sediment loading data from many monitoring

Table 7.1 Summary of input data and calibration parameters

Category	Input data	Data source
The stream network	Stream network, stream lengths, sub-catchment boundaries, sub-catchment areas	Automated catchment delineation based on a 50 m DEM, with modification flow diversions
Stream load data	Monitoring station	The suspended sediment flux at 31 stations from 1982 to 2010 was obtained from the National Land with Water Information monitoring network
Sediment source data	Developing land, forest land, agricultural land, and water land	Land-use data were downloaded from the Ministry of Land, Infrastructure, Transport and Tourism, Japan, 2006
Environmental setting data	Mean annual precipitation	The 20-year (1990–2010) average from Japanese Meteorological Agency
	Catchment slope	Mean value of local slope, obtained from 50 m DEM
	Soil texture, soil permeability	Obtained from the 1:5.000.000-scale FAO/UNESCO Soil Map of the World and the National and Regional Planning Bureau, Japan
	Reservoir (dam) loss	The Japan Dam Foundation (http://damnet.or.jp/)

stations within the model boundary (dependent variable); (3) Sediment source data describing the source of all modeled sediments or other components (independent variables); (4) Environmental datasets describing the environmental settings of the modeled, which cause statistically in the amphibious transport of sediments (independent variables). The input datasets are described in more detail in Table 7.1.

7.4.2.1 The Stream Network

Figure 7.3 shows the hydrologic network and catchments used for the SPARROW model of the Ishikari River basin, which are obtained from a 50 m digital elevation model (DEM). In total, there were 900 stream reaches, each with an associated sub-catchment. River networks mainly include river arrival and sub-catchment characteristics, such as river length, water flow direction, reservoir characteristics (such as surface area), and local and total drainage areas. For example, the areas of the smallest sub-catchment are 0.009 km^2, and the largest one is 117 km^2. The median area for these sub-catchments is 15.9 km^2.

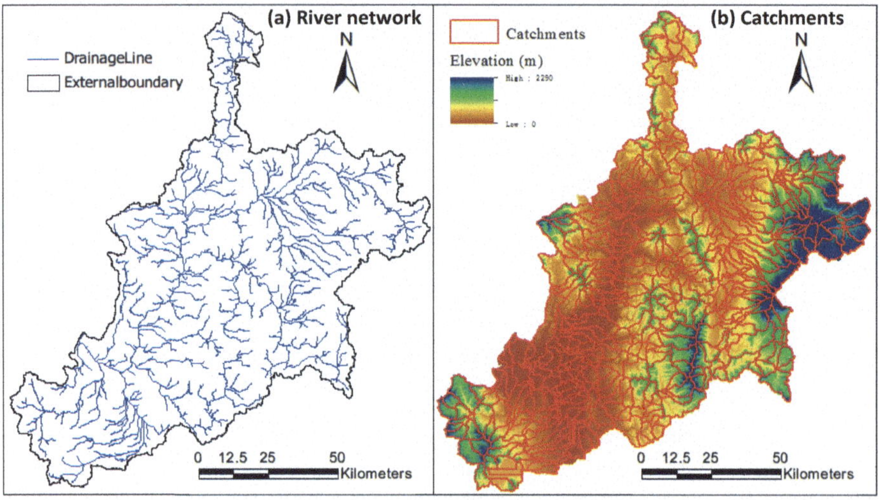

Fig. 7.3 Schematic showing **a** the river network and **b** 900 catchments in the Ishikari River Basin

7.4.2.2 Stream Load Data

Thirty-one stations (see Fig. 7.1) were selected for model calibration. The SS concentration and daily flow data for each station from 1982 to 2010 were collected through the National Land Water Information (http://www1.river.go.jp/) monitoring network. However, some flow measurement stations have very short recording times or lack of flow values, reflecting the gap in long-term monitoring. A streamflow record extension method called the Maintenance of Variance-Extension type 3 (MOVE.3) (Vogel and Stedinger 1985) was applied to estimate missing flow values or to extend the record at a short-record station based on daily streamflow values recorded at nearby, hydrologically similar index stations. On this basis, the FORTRAN Load Estimator (LOADEST) was used to estimate SS loads.

According to the lowest Akaike information criterion (AIC) values, some of the regression equations did not contain all terms (Sakamoto et al. 1986). The long-term average annual load of 31 stations was then standardized to the 2006 base year. The 2006 base year was chosen to be consistent with the latest explanatory geospatial data. Figure 7.4 shows the observed water flows (m^3/s) and the observed SS concentration (mg/l) at 31 monitoring stations in the Ishikari River Basin.

7.4.2.3 Sediment Source Data

Suspended sediment sources can be divided into sediments in highland areas, sediments in urban areas, and sediment erosion in river corridors (Langland et al. 2003). It is generally believed that the effects of land use lead to increased sediment loading and are therefore unintended consequences of human activities. In addition, land

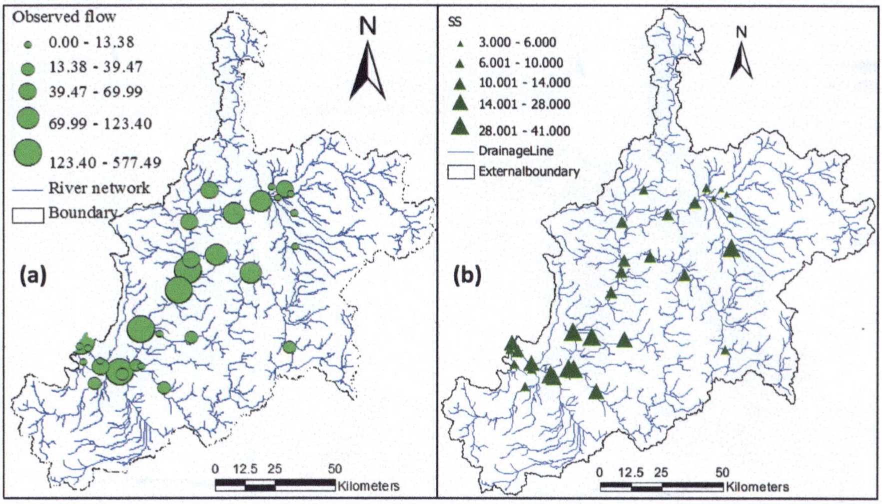

Fig. 7.4 Schematic showing **a** the observed water flows (m^3/s) and **b** the observed SS concentration (mg/l) at 31 monitoring stations

use and land-use change are also important factors affecting erosion and sediment yield. For example, when large areas of roads, roofs, and parking lots are covered with impervious surfaces, urbanization may eventually lead to a reduction in local surface erosion rates (Wolman 1967). The soil surface is increased due to the removal of natural plant cover, and agricultural land can greatly accelerate the erosion rate (Lal 2001). In addition, stream channel erosion may be the main source of sediment production in urbanized areas (Trimble 1997).

The SS source variables tested in the Ishikari SPARROW model include estimates of development sites, forest land, agricultural land, and river channels. Figure 7.5a shows 11 types of original land use in Ishikari River Basin, which were developed using data derived from the Policy Bureau of the Ministry of Land, Infrastructure, Transport and Tourism, Japan, 2006. They were then combined into four types (see Fig. 7.5b): developing land, forest land, agricultural land, and water land. Finally, different lands are allocated to individual sub-catchments to obtain the proportion for each land use using GIS zonal processes.

7.4.2.4 Environmental Setting Data

Climatic and landscape characteristics can affect SS transportation, which mainly contain climate, topography, and soil (Asselman et al. 2003; Dedkov and Mozzherin 1992). In this chapter, temperature, precipitation, slope, and soil permeability are used to assess the impact of "land-to-water" transport conditions. Mean annual temperature and precipitation data, representing the 20-year (1990–2010) average, were downloaded from the Japan Meteorological Agency. The results of them are

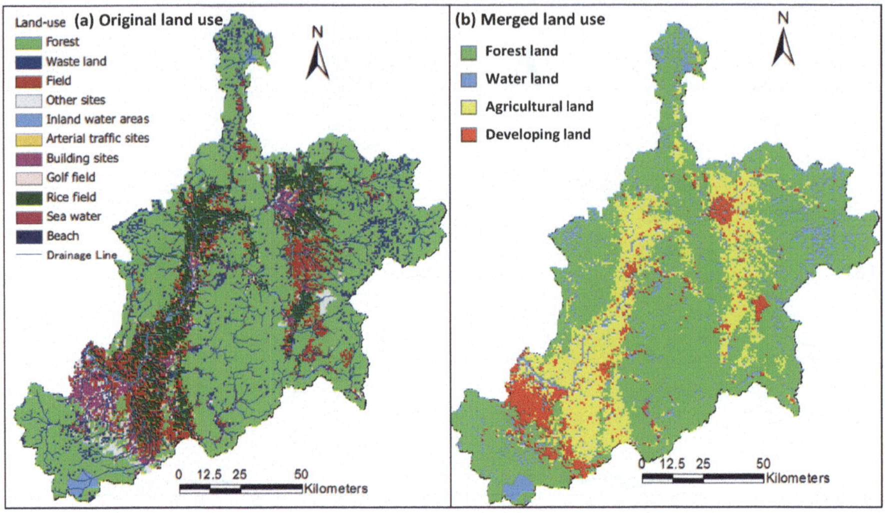

Fig. 7.5 Original and merged land use of the Ishikari River Basin, 2006

shown in Fig. 7.6a, b, which indicate that temperature ranged from 4.69 to 8.69 °C, and precipitation ranged from 1573.05 to 3013.57 mm. Figure 7.6c shows the slope of the basin, which is calculated using the GIS surface tool. Figure 7.6d shows the soil permeability and clay content, which are estimated using the data derived from the 1:5.000.000-scale FAO/UNESCO Soil Map of the World (Fao 1988) and the National and Regional Planning Bureau, Japan.

Loss of reaches and reservoirs are used as the mediating factors to influence the transfer of sediment from the river network. Reach-loss variable is nonzero only for stream reaches, and is defined for two separate categories including shallow-flowing (small) streams and deep-flowing (large) streams. Since the depth of the water flow is unknown, it is assumed that the water flow with a drainage area of less than 200 km^2 is a shallow and small streams. The loss of the reservoir is expressed in terms of the areal hydraulic load of the reservoir, which is calculated from the quotient of the average annual water storage outflow and surface area (Hoos and McMahon 2009).

7.5 Results

7.5.1 Streamflow Extension

Based on the MOVE.3 method and nearby long-time gauged stations, these stations with short recording records or missing flow values were filled and extended. Several criteria were used when selecting the long-term index stations (Nielsen 1999).

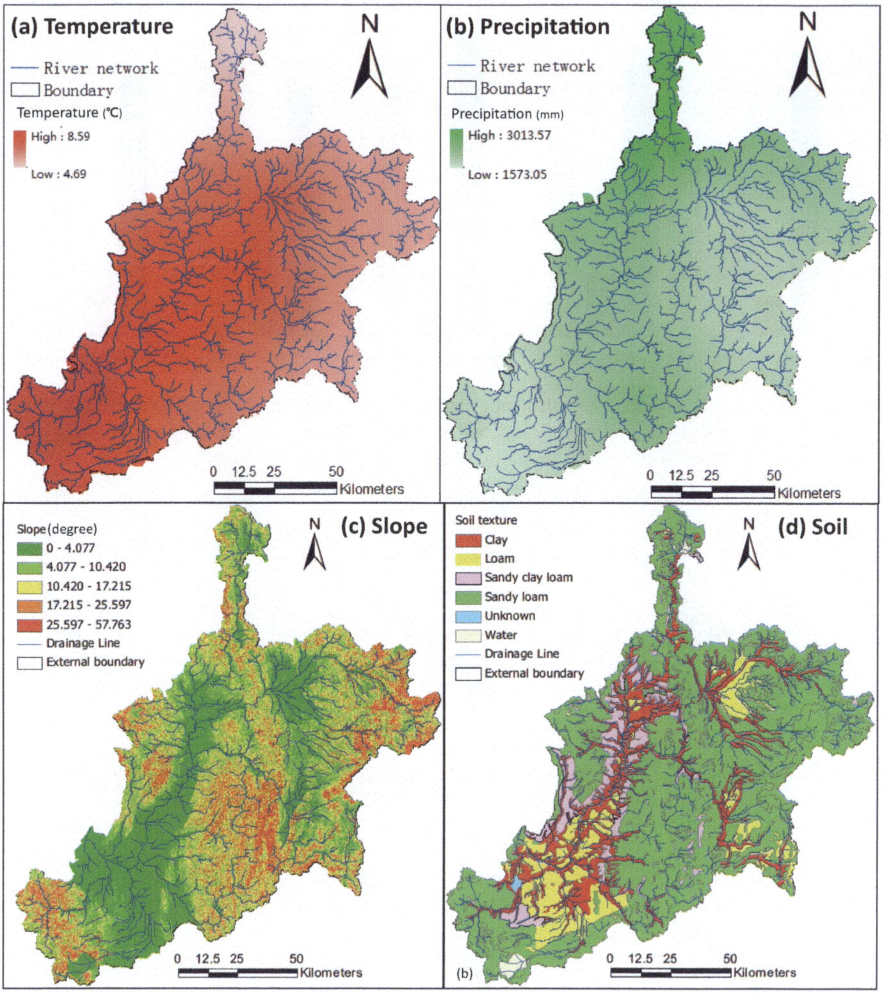

Fig. 7.6 Schematic showing the **a** temperature (°C), **b** precipitation (mm), **c** slope (degree), and **d** soil texture in Ishikari River Basin

Figure 7.7 shows Pearson's correlation coefficient (ρ), ranging from 0.79 (Umaoyibashi) to 0.96 (Yiwamizawa-oohashi), with an average value of 0.88, which suggests that relationship between the index stations and the studied stations were strong. The scatterplots of simultaneous average daily loads between the study stations and nearby index stations are shown in Fig. 7.8, indicating Yiwamizawa-oohashi station which had the best performance (see Fig. 7.8e).

After determining the index stations, the MOVE.3 method was used to extend the daily discharge records for each study station. During this process, two statistics including the root mean square error (RMSE) and the Nash–Sutcliffe efficiency

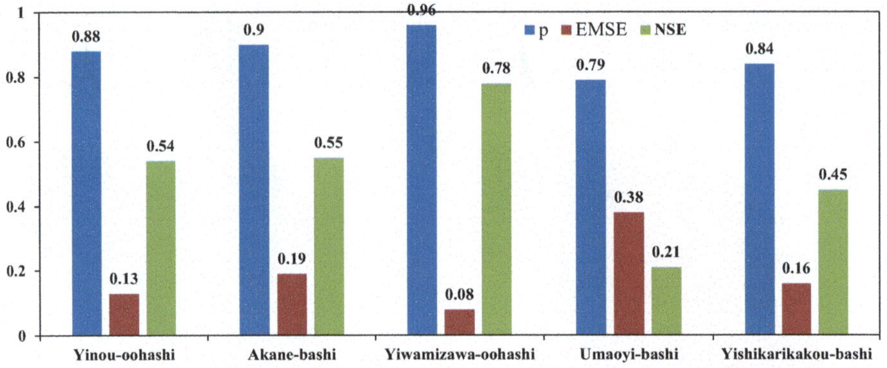

Fig. 7.7 Results of Pearson's correlation coefficient ρ for correlation between logarithms of river discharges at study stations and nearby index stations, RMSE of the estimating equations at each index station, and NSE of the model performance in river discharge simulations

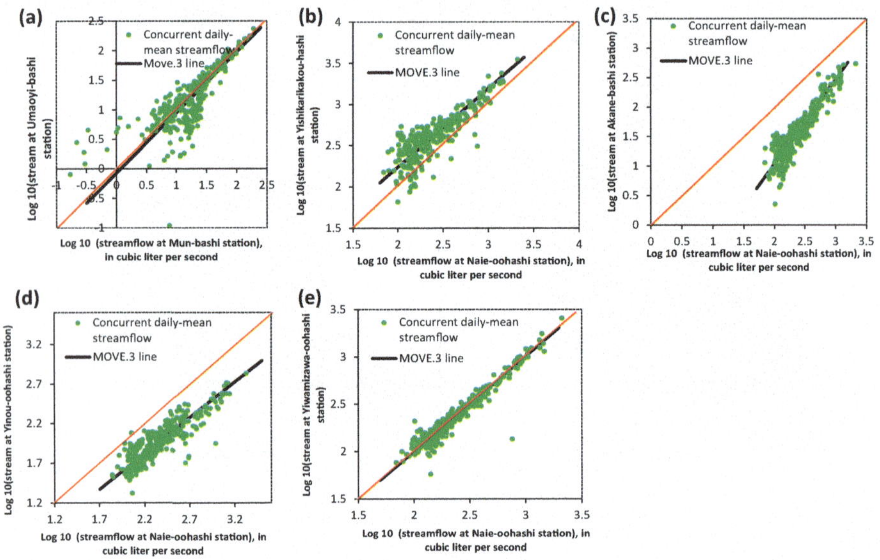

Fig. 7.8 The scatterplots of concurrent daily mean discharge between study stations and nearby index stations. The red line is the 1:1 line

(NSE, Legates, and McCabe 1999) were employed to evaluate the accuracy of the MOVE.3 method. The results were shown in Fig. 7.2. The RMSE ranged from 0.08 (Yiwamizawa-oohashi) to 0.38 (Umaoyi-bashi) and the NSE coefficient ranged from 0.21 (Umaoyi-bashi) to 0.78 (Yiwamizawa-oohashi), with an average of 0.51. Of them, Umaoyi-bashi had the largest RMSE and the smallest NSE, suggesting that the model performance of the site was relatively poor.

7.5.2 Regression Evaluation

Table 7.2 shows the regression coefficients, coefficients of determination (R^2), and AIC for load models used to estimate TN, TP, and SS at five sites in the Ishikari River Basin, Japan, 2000–2010. As can be seen in the figure, R^2 ranged from 71.86 to 90.94%, which indicates that the best-fit regression models for loads of TN, TP, and SS for the five studied sites performed well. Of them, Akane-bashi station was the best.

Meanwhile, the lowest AIC was the criterion to determine the best model and the AMLE was used to calculate the coefficients. For TN estimation at site Akane-bashi, the lowest of AIC was 0.729 and the coefficients *a*, *b*, *c*, *d*, *e*, *f*, and *g* were 7.694, 1.015, −0.066, 0.234, −0.040, and −0.017; while at site Umaoyi-bashi, the lowest of AIC was 0.928 and the coefficients employed were *a* (6. 354), *b* (0. 986), and *c* (0. 021). For TP estimation at site Akane-bashi, the lowest of AIC was 1.359 and the coefficients *a*, *b*, *c*, *d*, *e*, *f*, and *g* were 5.285, 1.064, −0.015, 0.044, –0.566, −0.024, and −0.002; while at site Yiwamizawa-oohashi, the lowest of AIC was 0.972 and the coefficients employed were *a* (7.445), *b* (1.286), *c* (0.095), *d* (−0.195), and *e* (−0.014). For SS estimation at site Akane-bashi, the lowest of AIC was 2.247 and the coefficients *a*, *b*, *c*, *d*, *e*, *f*, and *g* were 11.305, 1.597, 0.012, −0.051, −0.390, −0.040, −0.003, and 84.50; while at site Umaoyi-bashi, the lowest of AIC was 2.432 and the coefficients employed were *a* (10.103), *b* (1.592), *c* (0.138), *d* (0.363), *e* (0.348), *f* (−0.026), and *g* (−0.002).

7.5.3 Estimated Loads

This study mainly discussed the monthly, seasonal, and annual load estimates. For the sake of brevity, only estimated monthly average and measured instantaneous loads (kg/day) of TN, TP, and SS at site Akanebashi, 1985–2010, are displayed in time-series graphs (Fig. 7.9). As a measure of error associated with monthly average load estimates, the upper and lower 95% confidence intervals are also presented in Fig. 7.9. Yishikarikakou-bashi had the largest estimates, with TN, TP, and SS loads ranging from 8519.00 to 200189.00 kg/day (April 1999), 395.87 to 52299.00 kg/day (April 1999), and 92111.00–92500000.00 kg/day (September 2001), respectively. At all sites, monthly average TN, TP, and SS loads displayed seasonal fluctuations in both loads and in discharge from 1985 to 2010, even though the dates of peak discharge were not the same every year.

Figure 7.10 shows the estimated average loads of TN, TP, and SS for all stations from 1985 to 2010, which are averaged by the monthly averages for each month of the year (for example, the monthly average for all January, the average of February, the average of March, and so on). As can be seen from Fig. 7.10, April had the largest estimated average loads for TN, TP, and SS for all stations, the values of which are been in shown Fig. 7.6. After April, the estimated average load

Table 7.2 Regression coefficients, coefficients of determination (R^2), and AIC for load models used to estimate TN, TP, and SS at five sites in the Ishikari River Basin, Japan, 2000–2010

Water quality	Station's name	Regression coefficient							R^2 (%)	AIC
		a	b	c	d	e	f	g		
TN	Akane-bashi	7.694	1.015	−0.066	0.234	−0.040	−0.017		90.32	0.729
	Yinou-oohashi	9.061	0.973	0.017	0.178	0.185	−0.012	0.001	83.39	0.096
	Yiwamizawa-oohashi	10.144	0.932	0.149	0.125				90.92	−0.405
	Yishikarikakou-bashi	10.718	0.862	−0.074	−0.150	0.104	−0.011	0.001	81.61	0.123
	Umaoyi-bashi	6.354	0.986	0.021					89.22	0.928
TP	Akane-bashi	5.285	1.064	−0.015	0.044	−0.566	−0.024	−0.002	87.67	1.359
	Yinou-oohashi	6.039	0.963	0.103	0.185	−0.086	−0.014		71.86	1.078
	Yiwamizawa-oohashi	7.445	1.286	0.095	−0.004	−0.195	−0.014		85.29	0.972
	Yishikarikakou-bashi	7.886	1.164	0.101	0.150	0.031	−0.008	0.001	88.71	0.368
	Umaoyi-bashi	3.751	1.246	0.069	0.175	0.385	−0.017	−0.002	90.94	1.339
SS	Akane-bashi	11.305	1.597	0.012	−0.051	−0.390	−0.040	−0.003	84.50	2.247
	Yinou-oohashi	11.807	1.522	0.321	0.121	−0.117	−0.011	−0.003	75.68	1.777
	Yiwamizawa-oohashi	13.641	1.972	0.100	−0.124	−0.209	−0.029	−0.002	85.98	1.752
	Yishikarikakou-bashi	13.283	2.089	0.305	0.223	−0.141	−0.033		85.58	1.815
	Umaoyi-bashi	10.103	1.592	0.138	0.363	0.348	−0.026	−0.002	84.21	2.432

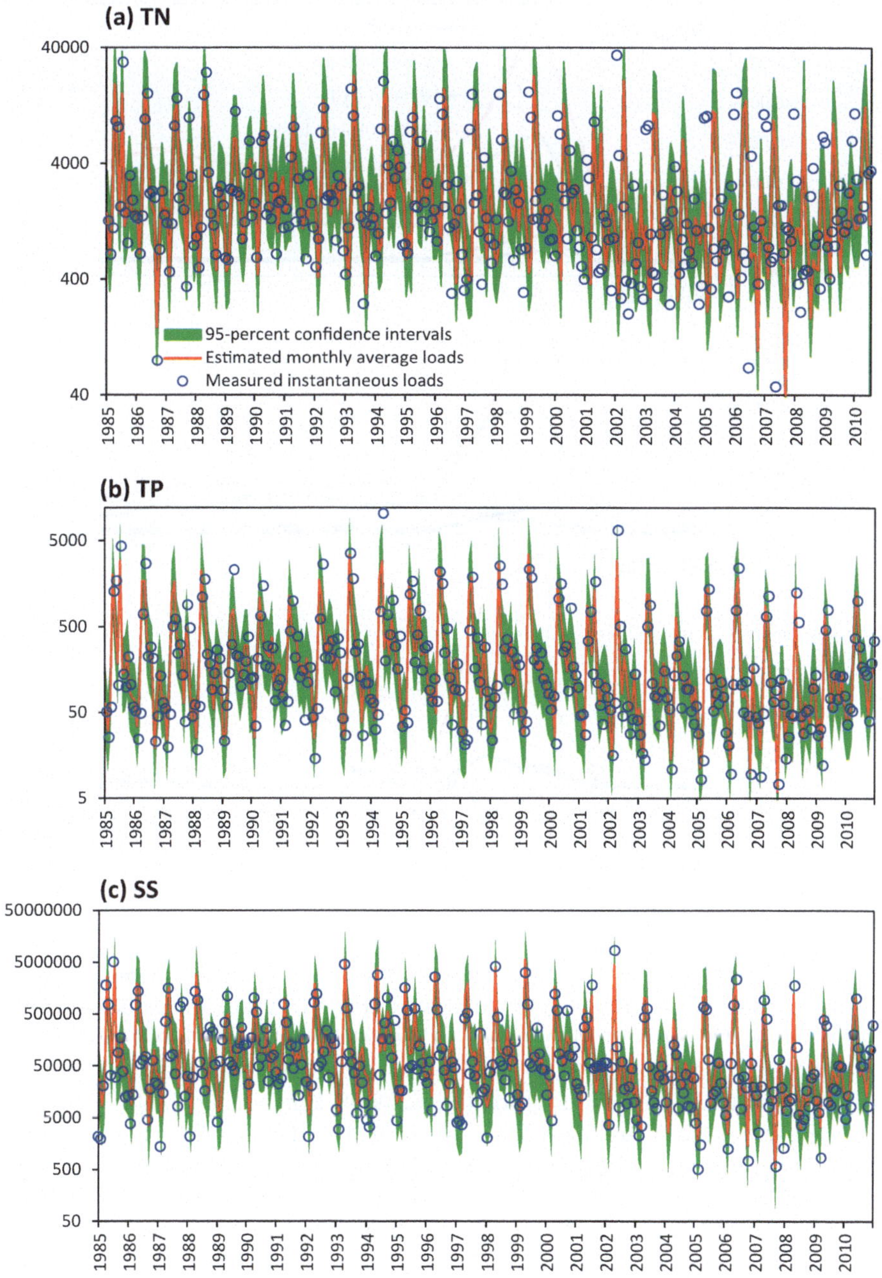

Fig. 7.9 Estimated monthly average and measured instantaneous loads (kg/day) of TN, TP, and SS at site Akanebashi, 1985–2010

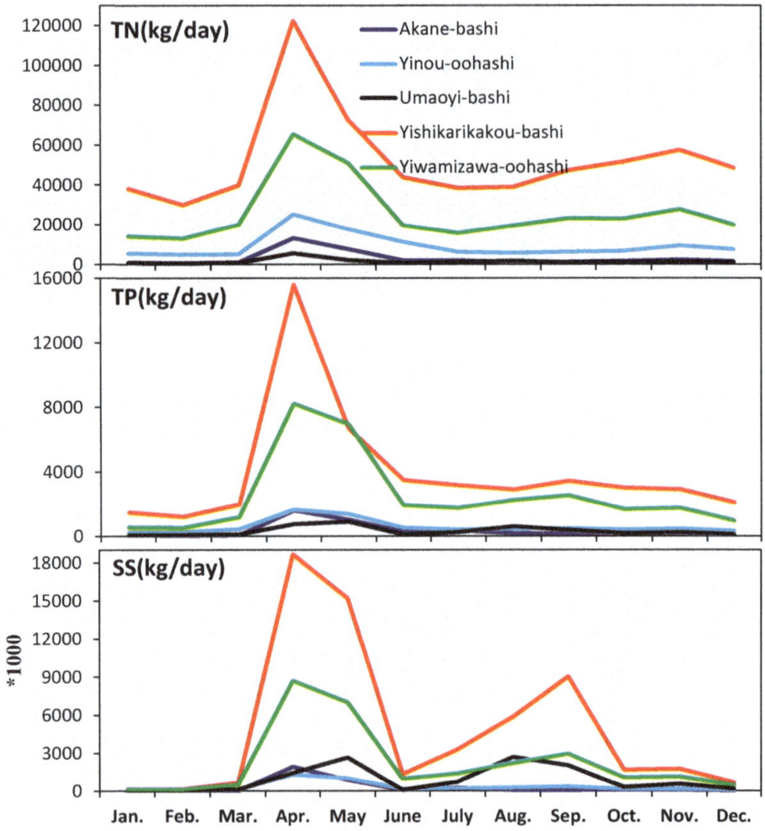

Fig. 7.10 Estimated monthly loads of TN, TP, and SS at five sites on the Ishikari River, from the January 1985 to the December 2010

drops and then increases. Typically, TP and SS peak in September, while loads in January, February, and March were relatively low. At Yishikarikakou-bashi, the values of TN, TP, and SS estimates were 122629.31 kg/day, 15630.62 kg/day, and 18694484.69 kg/day, respectively; at Yiwamizawa-bashi, the values of TN, TP, and SS estimates were 65651.08 kg/day, 8259.08 kg/day, and 8741123 kg/day, respectively; and at Yinou-bashi, the values of TN, TP, and SS estimates were 25375.77 kg/day, 1668.73 kg/day, and 1321272 kg/day, respectively.

Estimated seasonal loads of TN, TP, and SS at five sites were highly variable between 1985 and 2010 in the Ishikari River and its tributaries, with the greatest loads occurring in the spring and the smallest loads occurring in the winter (Fig. 7.10), reflecting fluctuations in discharge as a result of the combined effects of seasonal runoff patterns, the exact timing of which vary from year to year. At site Akane-bashi, TN load decreased from 8146.00 kg/day in spring to 1191.00 kg/day in winter, TP load decreased from 907.00 to 60.30 kg/day, and

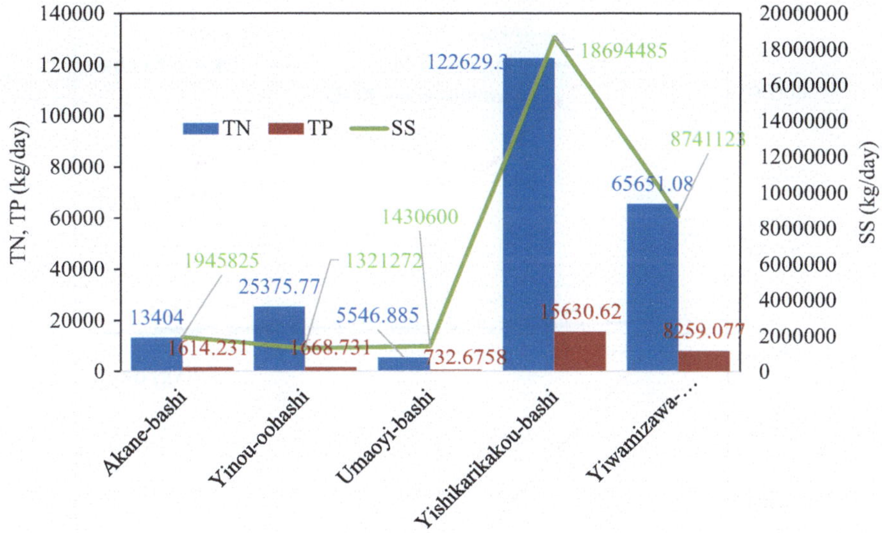

Fig. 7.11 Estimated average loads of TN, TP, and SS in April, at five sites on the Ishikari River, January 1985 through December 2010 (kg/day)

SS decreased from 956854.00 to 26906.00 kg/day. At site Yishikarikakou-bashi, TN load decreased from 78478.00 kg/day in spring to 39091.00 kg/day in winter, TP load decreased from 8110.00 to 1619.00 kg/day, and SS decreased from 11470000.00 to 367458.00 kg/day. Seasonal fluctuations were consistent with monthly fluctuations (Fig. 7.10). Regardless of season, site Yishikarikakou-bashi had the largest loads of TN, TP, and SS, far more than at other sites, the seasonal mean of which were 46 702, 3 560, and 1 991 033 kg/day (Fig. 7.11).

Between 1985 and 2010, the estimated seasonal load of TN, TP, and SS at the five locations of the Ishikari River and its tributaries varied greatly, with the largest loads in spring and the least load in winter (Fig. 7.12), which were line with the seasonal runoff patterns of every year. At the Akane-bashi station, the total nitrogen load decreased from 8146.00 kg/day in spring to 1191.00 kg/day in winter, and the total phosphorus load decreased from 907.00 to 60.30 kg/day, while SS decreased from 956854.00 to 26906.00 kg/day. At the Yishikarikakou-bashi site, the TN load decreased from 78478.00 kg/day in spring to 39091.00 kg/day in winter, the TP load decreased from 8110.00 to 1619.00 kg/day, and the SS decreased from 11470000.00 to 367458.00 kg/day. At the Yiwamizawa-bashi site, the TN load decreased from 45481.00 kg/day in spring to 16067.00 kg/day in winter, the TP load decreased from 5449.00 to 712.86 kg/day, and the SS decreased from 5392.07 to 256.53 kg/day.

Seasonal fluctuations are consistent with monthly fluctuations. As can be seen in Fig. 7.13, the Yishikarikakou-bashi site had the highest seasonal TN, TP, and SS loads, far exceeding other sites, with seasonal averages of 46,702, 3,560, and 1,991,033 kg/day, followed by Yiwamizawa-oohashi site (TN(26233.50 kg/day), TP(2553.00 kg/day), and SS(2250293.00 kg/day)).

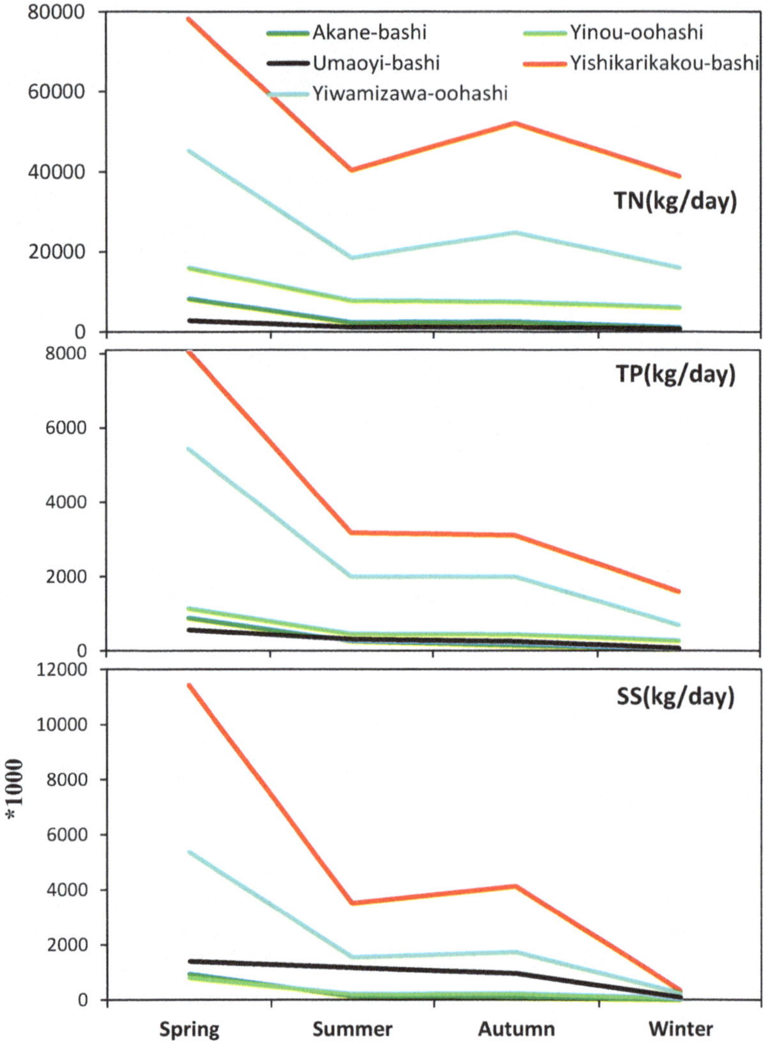

Fig. 7.12 Estimated seasonal average loads of TN, TP, and SS at five sites on the Ishikari River, January 1985 through December 2010

7.5.4 Trends of the Estimated Yearly Loads of TN, TP, and SS

Since the Yishikarikakou-bashi site is located on the inlet of the Ishikari River, this study calculated the trends of the estimated yearly loads of TN, TP, and SS at this station, which could reflect changes of water quality for the whole Ishikari River Basin. The results are shown in Fig. 7.14, which suggest that a significant decreasing trend was detected for all the water quality indices at Yishikarikakou-bashi site

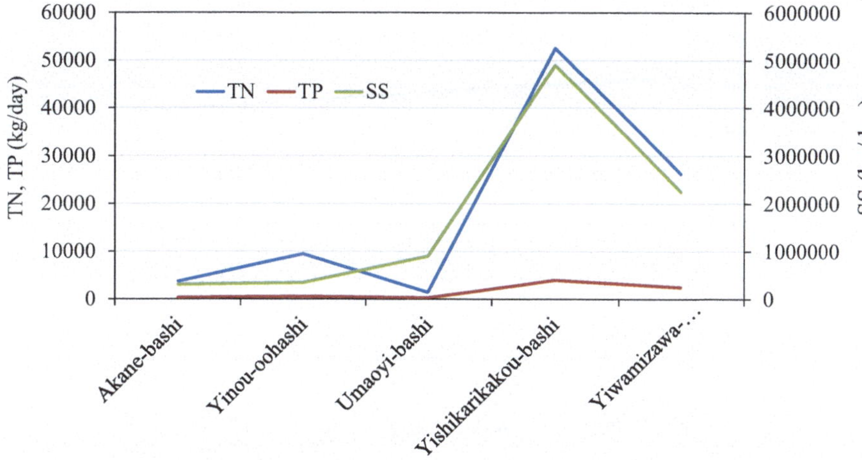

Fig. 7.13 Estimated seasonal loads of TN, TP, and SS at five sites on the Ishikari River (kg/day)

from 1985 to 2010. The Mann–Kendall Z values of TN, TP, and SS were −3.79, −2.42, and −2.82, respectively, and the decrease of TN, TP, and SS was about 635.04 kg/day, 65.68 kg/day, and 217257 kg/day for each year, respectively. Results show a significant improvement in water quality in Ishikari River Basin.

7.5.5 Results of SPARROW SS Model

7.5.5.1 Model Calibration

After parameter adjustment, we successfully established the final SPARROW suspended sediment model for the Ishikari River Basin, which contains four source variables (agricultural land, forest land, developing land, and streambed (stream channels)), three landscape transport variables (precipitation, slope, and soil permeability), two in-stream loss coefficients (small stream (drainage area ≤ 200 km^2) and large stream (drainage area > 200 km^2)), as well as reservoir attenuation describing removal/deposition in reservoirs (see Tables 7.1 and 7.3). Table 7.3 lists the model calibration results for the logarithmic transformation and the nonlinear least-squares estimation of the sum in Eq. (7.11), which explains about 95.96% (R^2) of the spatial variation in the natural logarithm of the average annual suspended sediments in the Ishikari River Basin. The mean square error (MSE) of 0.323 indicates that the SS predicted by the model matches the observed load very well. For comparison, the accuracy of the results is worse than other SPARROW models such as Waikato SPARROW model ($R^2 = 0.97$ and MSE $= 0.14$, calibrated using 37 stations) due to different parameters and the accuracy of calibration data (Alexander et al. 2002).

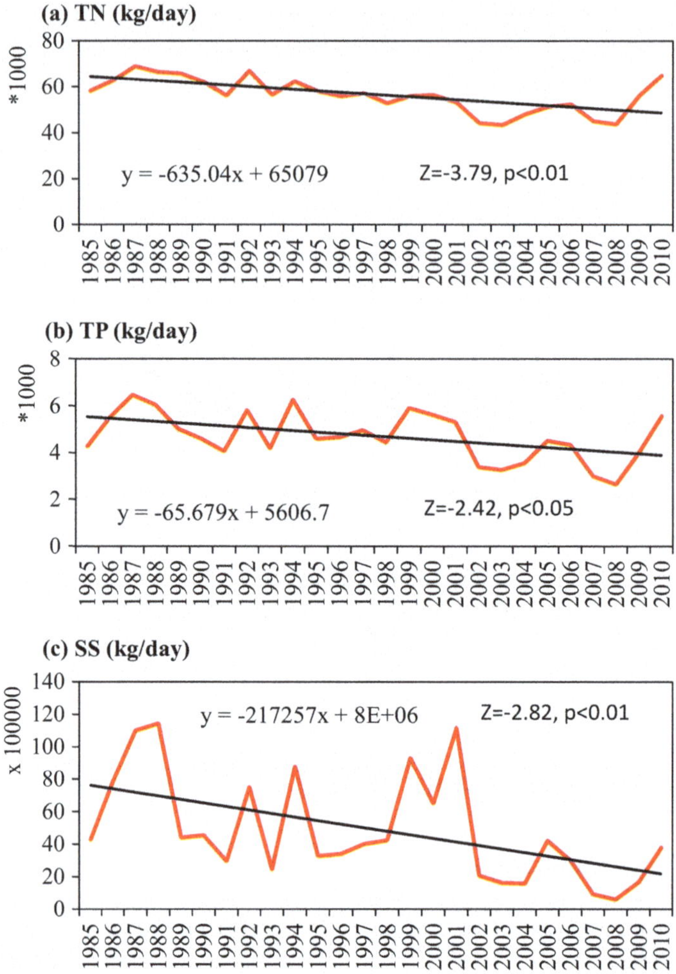

Fig. 7.14 Trends of estimated yearly loads of TN, TP, and SS at Yishikarikakou-bashi site on the Ishikari River (kg/day)

Figure 7.15a shows the observed and predicted SS flux (kg/year) at 31 monitoring sites included in the Ishikari SPARROW model (Natural logarithm transformation applied to observed and predicted values). As we can see from the figure, the fitting line is close to the red dashed line (1:1 line), suggesting that the predicted values were close to the observed values. Figure 7.15b shows the model residuals for 31 monitoring stations used to calibrate the final Ishikari SPARROW model which reveals that high-predicted (<0) monitoring sites mainly exist in the central part of the Ishikari River Basin, and high-predicted (>0) monitoring sites exist in the upper and lower river. Schwarz et al. 2006 argued that standardized residual greater than

Table 7.3 SPARROW estimates of model statistics for Ishikari River Basin SS based on the 31 monitoring stations

Model parameters	Coefficient units	Estimated coefficient	Standard error	P-value
SS sources				
Developing land	kg/km^2/year	1006.267	508.503	0.028
Forest land	kg/km^2/year	75.554	31.058	0.011
Agricultural land	kg/km^2/year	234.211	121.7511	0.036
Streambed (stream channels)	kg/km/year	123.327	99.567	0.113
Land-to-water loss coefficient				
Slope	–	0.349	0.094	<0.001
Soil permeability	h/cm	−9.195	2.431	<0.001
Precipitation	mm	0.007	0.002	<0.002
In-stream loss rate				
Small stream (drainage area ≤200 km^2)	day^{-1}	−0.044	0.011	<0.001
Big stream (drainage area >200 km^2)	day^{-1}	0.000012	0.0068	>0.050
Reservoir-loss	m/year	26.283	4.364	<0.001
Model diagnostics				
Mean square error	0.323			
Number of observations	31			
R-squared	0.9596			

Notes This table indicates the overall model calibration results, statistical parameter estimates, standard errors, and probability levels for modeled explanatory sediments variables. All sources and storage items are subject to non-negative estimates for more realistic physical simulation of the suspended sediment transport. Due to this specification, the statistical significance of the source and aquatic storage coefficient estimates is evaluated as a one-sided p statistic, and the two-sided values are used to assess the probability levels for land-to-water parameters (Schwarz et al. 2006)

3.6 are generally considered outliers and require further investigation. Therefore, the final model did not show evidence of large prediction bias at the monitoring site.

All simulated source variables were statistically significant ($P < 0.05$) except for the river channel and indicated the average level of sediment supply (Table 7.3). The developing land produced the largest intrinsic sediments, with an estimated value of approximately 1006.267 kg/km^2/year, which is line with the results from Brakebill et al. (2010) and Schwarz (2008). The sediment yield of agricultural land is the second highest, with an estimated value of about 234.211 kg/km^2/year, while the sediment yield of forest land is the lowest, with an estimated value of about 75.554 kg/km^2/year.

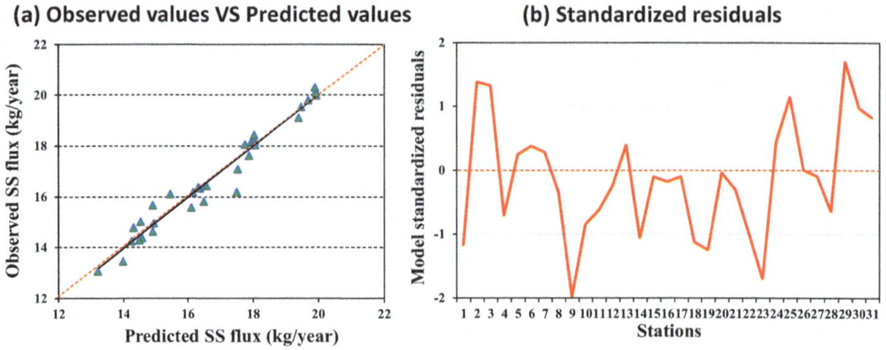

Fig. 7.15 **a** Observed and predicted SS flux (kg/year) at 31 monitoring sites included in the Ishikari SPARROW model (Natural logarithm transformation applied to observed and predicted values); **b** Model residuals for 31 monitoring stations used to calibrate the final Ishikari SPARROW model

Land-to-water delivery for sediment land sources is powerfully mediated by watershed slope, soil permeability, and rainfall, all of which are statistically significant (Table 7.3). As expected, Table 7.3 shows that sediment produced from land transport to rivers is most efficient in areas with greater basin slope, less permeable soils, and greater rainfall, which is consistent with the results calculated by Brakebill et al. (2010). The alteration of these factors can directly and indirectly cause changes in sediment degradation and deposition, and, finally, to the sediment yield (Luce and Black 1999; Nelson and Booth 2002). Increased rainfall amounts and intensities can directly increase surface runoff, leading to greater rates of soil erosion (Nearing et al. 2005; Ran et al. 2012) with consequences for productivity of farmland (Julien and Simons 1985). Watershed slope and soil permeability have a powerful influence on potential surface runoff as they affect the magnitude and rate of eroded sediment that may be transported to streams (Brakebill et al. 2010).

The slope of the watershed, soil permeability, and precipitation have an important impact on the suspended sediment transport from land to water. Table 7.3 shows that suspended sediments produced from land transport to rivers are most effective in regions with larger watershed slopes, less permeable soils, and greater precipitation, which is consistent with the calculations of Brakebill et al. (2010). Changes in these factors can directly and indirectly lead to changes in sediment degradation and sedimentation, and ultimately to changes in suspended sediments yield (Luce and Black 1999; Nelson and Booth 2002). For example, increased precipitation amounts and intensities can directly increase surface runoff, which can bring greater soil erosion rates (Nearing et al. 2005; Ran et al. 2012), affecting farmland productivity (Julien and Simons 1985).

Table 7.3 also shows that the estimated coefficient for large streams and small streams is about 0.000012 day^{-1} and −0.044 day^{-1}, respectively, which suggests that suspended sediments are removed from large streams and accumulated in small streams. These results are not in line with one previous point that the river channels with high flow can increase the amount of sediments generated from stream channels

(Schwarz 2008). In the reservoir, sediment storage was statistically significant and estimated to be approximately 26.283 m/year, which is similar to the result of the conterminous U.S. SPARROW model (36 m/year) (Schwarz 2008).

7.5.5.2 Model Application

Figure 7.16a shows the spatial distribution of total suspended sediments yields, ranging from 0.034 to 1190 kg/ha/year (mean = 101 kg/ha/year). The distribution also shows the total suspended sediments yields concentrated in the secondary watershed in the middle and lower reaches of the Shishou River, which suggests that the sediments yields relate to the river network. Figure 7.16b shows the spatial distribution of increased suspended sediments yields, which indicates that much of the incremental sediment yields are distributed in regions with high total suspended sediments yields. The largest value of the increased suspended sediments yields is greater than 150 kg/ha/year.

Figure 7.16 also illustrates the percentage of total incremental flux produced by (c) agricultural lands, (d) developing lands, (e) forested lands, and (f) stream channels, indicating the relative contribution of various sources to each sub-basin. As shown in the figure, the proportion of agricultural-land sediment yields was greater than 50%, in many sub-basins with high total sediment yields. The same distribution was found in the proportion of predicted forest-land sediment yield and predicted steam channels yield, and all these two sources produced low suspended sediments in the Ishikari River Basin. Figure 7.17 shows the total incremental flux generated for agricultural lands, developing lands, forested lands, and stream channels. As shown in the figure, agricultural lands produced the largest suspended sediments, up to 35.11%, followed by forested lands (23.42%), developing lands (22.91%), and stream channels (18.56%).

7.6 Discussions

7.6.1 Large Loads of TN, TP, and SS at Site Yishikarikakou-bashi

The Yishikarikakou-bashi site had the highest loads of TN, TP, and SS in both monthly and seasonal loads. The reason is that this site is located in the lower reaches of the Ishikari River and has the highest average discharge, which is the primary driver of constituent delivery to coastal waters (Fig. 7.18). Constituents from the upper, middle, and lower reaches of the mainstream and the Uryū and Yūbari Tributaries move together and are discharged at the Yishikarikakou-bashi Site.

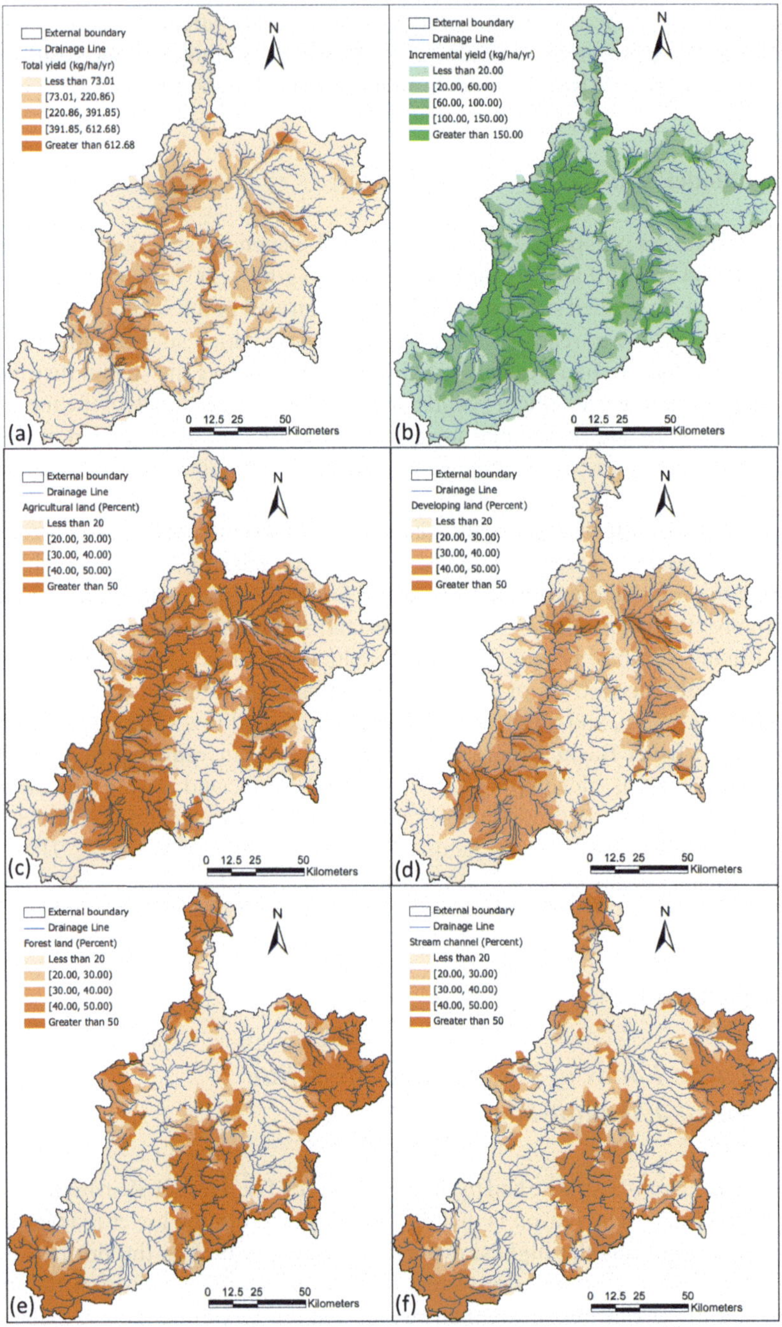

Fig. 7.16 Map showing the spatial distribution of **a** total suspended sediment yields, **b** incremental suspended sediment yields, **c** sediments produced by agricultural lands, **d** sediments produced by developing lands, **e** sediments produced by forest lands, and **f** sediments produced by stream channels

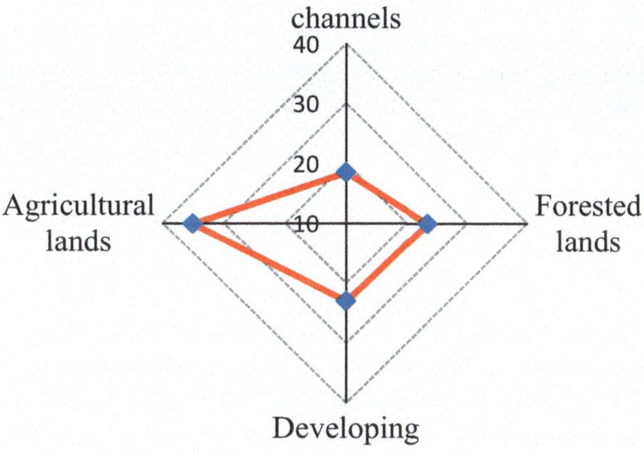

Fig. 7.17 The total incremental flux generated for agricultural lands, developing lands, forested lands, and stream channels

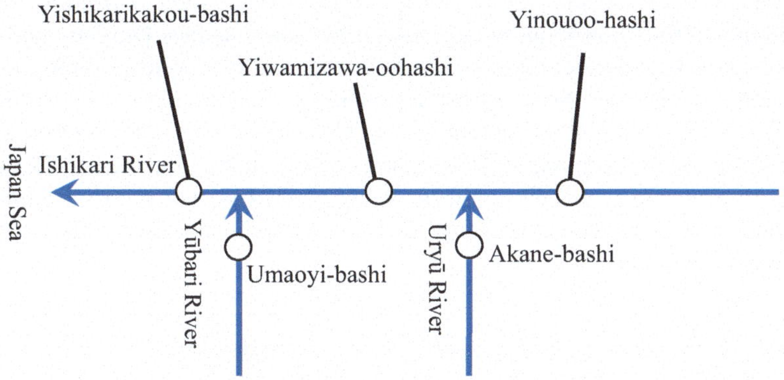

Fig. 7.18 Schematic diagram of the Ishikari River and five studied sites

7.6.2 Decreasing Trends of TN, TP, and SS Loads

From Fig. 7.9, we can see a downward trend of TN, TP, and SS load (most obvious was the decline of SS after 2001), which is in line with the result from Luo et al. (2011). Many researchers have shown that land use has a large impact on water quality and there is a significant correlation between water quality parameters and land-use types (Tu 2011; Woli et al. 2004). In Hokkaido, the urbanization process was accelerating, leading to an increase in the population of the Ishikari River Basin. Table 7.4 shows that the population of the Ishikari River Basin has increased from 760132 in 1920 to 3124348 in 2005, with 32.2% of Hokkaido's population in 1920 to 55.5% of Hokkaido's population in 2005. Urbanization has increased the pressure on the Ishikari River Basin, such as the increase of flooding, river erosion, and so on (Klein 2007).

However, numerous water management measures have been implemented in Japan to prevent deterioration of water quality. First of all, many polices and laws have been

Table 7.4 Increasing trend of people that live in the Ishikari River Basin from 1920 to 2005

Year	Hokkaido	Ishikari river Basin	Percentage (%)
1920	1599051	760132	32.2
1955	2942693	1830394	38.3
1965	3044706	2127094	41.4
1985	2876589	2802850	49.4
2005	2507785	3124348	55.5

enacted to reduce and mitigate pollution, especially since 2000 (see Fig. 7.19), leading to a comprehensive policy framework concerning water pollution control. In the 1990s, "Basic Environment Law" was enacted to set environmental quality standards for water quality. "Drinking Water Sources Law", "Laws Concerning Special Measures for Conservation of Drinking Water Sources", and "Law on Livestock Excreta Management and Recycling" were enacted to keep water sources from being polluted. In the 2000s, "Law Concerning Special Measures against Dioxins" and "Soil Contamination Countermeasures Law" were enacted to prevent pollution caused by hazardous substances. Some important international conventions concerning the prevention of water pollution including the "Convention on the Prevention and of Marine Pollution by Dumping of Wastes and other Matter", "International Convention for the Prevention of Pollution from Ships", and "International Convention for the Control and Management of Ships' Ballast and Sediments" were also enforced to protect water quality.

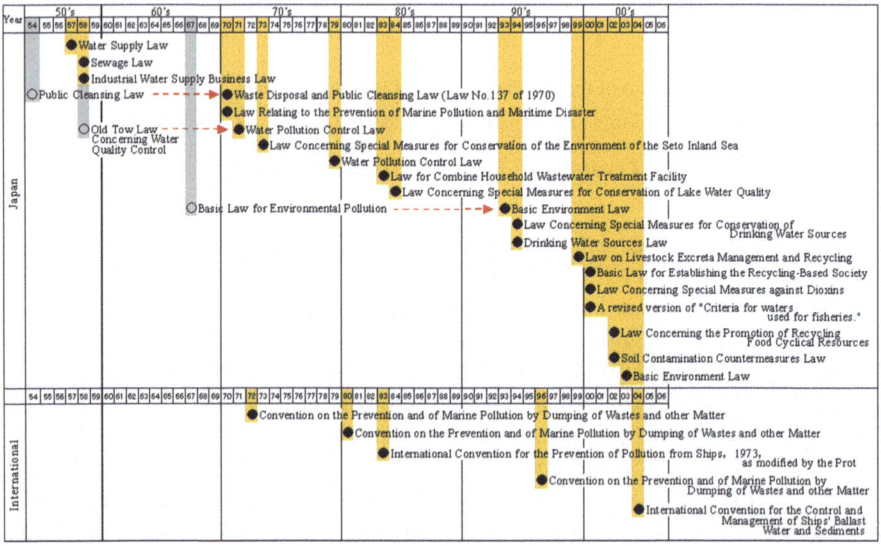

Fig. 7.19 A History of enactment of laws concerning water pollution control (*Source* https://www.jetro.go.jp/ttppoas/special/env_rep_english/env_rep_03_1.html)

However, Japan has taken many water management measures to prevent water quality deterioration. First, especially since 2000, many policies and laws to reduce and mitigate pollution have been enacted (see Fig. 7.19), resulting in a comprehensive policy framework for water pollution control. In the 1990s, the "Basic Environment Law" was enacted to set environmental quality standards for water quality for whole Japan. "Drinking Water Sources Law", "Laws Concerning Special Measures for Conservation of Drinking Water Sources", and "Law on Livestock Excreta Management and Recycling" were promulgated to prevent pollution of water sources. In the 2000s, the "Law Concerning Special Measures against Dioxins" and "Soil Contamination Countermeasures Law" were promulgated to prevent pollution caused by harmful substances. Some important international conventions on the prevention of water pollution were also implemented.

In addition, with the urbanization of Japan, the water infrastructure including the sewer system has also expanded. As shown in Table 7.5, the length, processing area, processing capacity, and coverage of the sewer system in Sapporo City had increased from 910.30 km, 1771 ha, 115.4×10^3 m^3/day, and 19.20% in 1970 increased to 8139.70 km, 24611 ha, 1173.80×10^3 m^3/day, and 99.70% in 2009, respectively. Meanwhile, improvements in industrial water treatment and wastewater treatment systems and improvements in "Jukasou" (Japan's domestic wastewater treatment system) had also helped to reduce wastewater flow (Luo et al. 2011). Overall, improvements in policies and laws and improvements in water infrastructure in recent years have led to a downward trend in TN, TP, and SS loads.

Table 7.5 Changes of length of sewer system, processing area, processing capacity, and coverage rate in Sapporo (Data from https://www.city.sapporo.jp/kurashi/suido_gesui/index.html)

Year	Length of sewer system (km)	Processing area (ha)	Processing capacity (thousands of m^3/day)	Coverage rate of sewer system (%)
1970	910.30	1,771	115.4	19.20
1975	2,437.40	8,552	423	64.50
1980	4,170.30	14,638	729	85.90
1985	5,887.20	18,786	948.1	91.80
1990	6,753.70	20,602	986.8	95.40
1995	7,348.20	22,933	1,044.80	98.60
2000	7,713.70	23,813	1,089.80	99.20
2005	8,006.60	24,402	1,173.80	99.50
2009	8,139.70	24,611	1,173.80	99.70

7.7 Conclusions

In this chapter, we firstly estimated the total nitrogen (TN), total phosphorus (TP), and suspended sediment (SS) loads in Ishikari River Basin, from January 1985 to December 2010, and then successfully established the final SPARROW suspended sediment model for the Ishikari River basin, which contains four source variables (agricultural land, forest land, developing land, and streambed (stream channels)), three landscape transport variables (precipitation, slope, and soil permeability), two in-stream loss coefficients (small stream (drainage area $\leq 200\ km^2$) and large stream (drainage area $> 200\ km^2$)), as well as reservoir attenuation describing removal/deposition in reservoirs. The significant conclusions of the calibration procedure and model application are summarized below:

(1) The loads of TN, TP, and SS were largest in April, and spring, reflecting the fluctuation of seasonal runoff patterns in the Ishikari River Basin. Also, the Yishikarikakou-bashi site had the highest water quality loads due to which it is located in the lower reaches of the Ishikari River.
(2) A decreasing trend was found in TN, TP, and SS loads in Ishikari River Basin from 1985 to 2010, because of improvements in policies and laws and improvements in water infrastructure in recent years.
(3) The calibration results of SPARROW model account for approximately 95.96% of the spatial variation in the natural logarithm of mean annual SS flux (kg/year) and show relatively small prediction errors based on 31 monitoring stations.
(4) Developing land has the largest intrinsic sediment yield at around 1006.267 $kg/km^2/year$, followed by agricultural land (234.211 $kg/km^2/year$), stream channels (123.327 $kg/km^2/year$), and forest land (75.554 $kg/km^2/year$). Reservoir attenuation (26.283 m/year) is statistically significant, indicating that reservoirs can play an important role in sediment retention.
(5) The total sediment yield and incremental production are concentrated in the middle and lower reaches of the Ishikari River, indicating that these sub-basins are most susceptible to erosion. The percentages of total incremental fluxes from agricultural land, developing land, forest land, and river channels were 35.11%, 23.42%, 22.91%, and 18.56%, respectively.

References

Alexander, R. B., Elliott, A. H., Shankar, U., & McBride, G. B. (2002). Estimating the sources and transport of nutrients in the Waikato River Basin, New Zealand. *Water Resources Research, 38,* 1268.

Alexander, R. B., Smith, R. A., & Schwarz, G. E. (2000). Effect of stream channel size on the delivery of nitrogen to the Gulf of Mexico. *Nature, 403,* 758–761.

Alexander, R. B., Smith, R. A., Schwarz, G. E., Boyer, E. W., Nolan, J. V., & Brakebill, J. W. (2007). Differences in phosphorus and nitrogen delivery to the Gulf of Mexico from the Mississippi River Basin. *Environmental Science and Technology, 42,* 822–830.

Armour, J. D., Hateley, L. R., & Pitt, G. L. (2009). Catchment modelling of sediment, nitrogen and phosphorus nutrient loads with SedNet/ANNEX in the Tully–Murray basin. *Marine & Freshwater Research, 60,* 1091–1096.

Asahi, K., Kato, K., & Shimizu, Y. (2003). Estimation of sediment discharge taking into account tributaries to the Ishikari River. *Journal of Natural Disaster Science, 25,* 17–22.

Asselman, N. E. M., Middelkoop, H., & Van Dijk, P. M. (2003). The impact of changes in climate and land use on soil erosion, transport and deposition of suspended sediment in the River Rhine. *Hydrological Processes, 17,* 3225–3244.

Beasley, D. B., Huggins, L. F., & Monke, E. J. (1980). ANSWERS: A model for watershed planning. *Transactions of the ASAE, 23.*

Brakebill, J. W., Ator, S. W., & Schwarz, G. E. (2010). Sources of Suspended-Sediment flux in streams of the chesapeake bay watershed: A regional application of the SPARROW model1. *JAWRA Journal of the American Water Resources Association, 46,* 757–776.

Carpenter, S. R., Caraco, N. F., Correll, D. L., Howarth, R. W., Sharpley, A. N., & Smith, V. H. (1998). Nonpoint pollution of surface waters with phosphorus and nitrogen. *Ecological Applications, 8,* 559–568.

Christensen, V. G., Jian, X., Ziegler, A. C., & Demonstration, E. B. G. R. (2000). *Regression analysis and real-time water-quality monitoring to estimate constituent concentrations, loads, and yields in the Little Arkansas River, south-central Kansas, 1995–99.* US Department of the Interior, US Geological Survey.

Dedkov, A. P., & Mozzherin, V. I. (1992). Erosion and sediment yield in mountain regions of the world. In *Erosion, debris flows and environment in mountain regions*, no. 209 (pp. 29–36).

Duan, W., He, B., Sahu, N., Luo, P., Nover, D., Hu, M., et al. (2017). Spatiotemporal variability of Hokkaido's seasonal precipitation in recent decades and connection to water vapour flux. *International Journal of Climatology, 37,* 3660–3673.

Duan, W., Takara, K., He, B., Luo, P., Nover, D., & Yamashiki, Y. (2013). Spatial and temporal trends in estimates of nutrient and suspended sediment loads in the Ishikari River, Japan, 1985 to 2010. *Science of the Total Environment, 461,* 499–508.

Duan, W. L., He, B., Takara, K., Luo, P. P., Nover, D., & Hu, M. C. (2015). Modeling suspended sediment sources and transport in the Ishikari River basin, Japan, using SPARROW. *Hydrology and Earth System Sciences, 19,* 1295–1306.

Duan, W. L., He, B., Takara, K., Luo, P. P., & Yamashiki, Y. (2012). Estimating the sources and transport of nitrogen pollution in the Ishikari River Basin, Japan. In *Advanced materials research* (pp. 3007–3010).

FAO. (1988). Soil map of the world, FAO/Unesco/ISRIC. Retrieved May 2019, from http://www.fao.org/soils-portal/soil-survey/soil-maps-and-databases/faounesco-soil-map-of-the-world/en/.

Ferguson, R. I. (1986). River loads underestimated by rating curves. *Water Resources Research, 22,* 74–76.

Gilbert, R. O. (1987). *Statistical methods for environmental pollution monitoring.* New York, NY: Van Nostrand Reinhold.

Gruber, N., & Galloway, J. N. (2008). An Earth-system perspective of the global nitrogen cycle. *Nature, 451,* 293–296.

Helsel, D. R., & Hirsch, R. M. (1992). *Statistical methods in water resources.* Elsevier Science.

Hoos, A. B., & McMahon, G. (2009). Spatial analysis of instream nitrogen loads and factors controlling nitrogen delivery to streams in the southeastern United States using spatially referenced regression on watershed attributes (SPARROW) and regional classification frameworks. *Hydrological Processes, 23,* 2275–2294.

Ishida, T., Nakayama, K., Okada, T., Maruya, Y., Onishi, K., & Omori, M. (2010). Suspended sediment transport in a river basin estimated by chemical composition analysis. *Hydrological Research Letters, 4,* 55–59.

Jeon, J. H., Lim, K. J., Yoon, C. G., & Engel, B. A. (2011). Multiple segmented reaches per subwatershed modeling approach for improving HSPF-Paddy water quality simulation. *Paddy and Water Environment, 9,* 193–205.

Johanson, R. C., Imhoff, J. C., & Davis, H. H. (1980). *Users manual for hydrological simulation program-FORTRAN (HSPF)*. Environmental Research Laboratory, Office of Research and Development, US Environmental Protection Agency.

Johnes, P. J. (2007). Uncertainties in annual riverine phosphorus load estimation: Impact of load estimation methodology, sampling frequency, baseflow index and catchment population density. *Journal of Hydrology, 332,* 241–258.

Julien, P. Y., & Simons, D. B. (1985). Sediment transport capacity of overland flow. *Transactions of the ASAE, 28,* 755–762.

Kendall, M. G. (1975). *Rank correlation methods*. London: Charles Griffin.

Kirsch, K., Kirsch, A., & Arnold, J. G. (2002). *Predicting sediment and phosphorus loads in the Rock River basin using SWAT*. Forest 971, 10.

Klein, R. D. (2007). Urbanization and stream quality impairment1. *JAWRA Journal of the American Water Resources Association, 15,* 948–963.

Kulasova, A., Smith, P. J., Beven, K. J., Blazkova, S. D., & Hlavacek, J. (2012). A method of computing uncertain nitrogen and phosphorus loads in a small stream from an agricultural catchment using continuous monitoring data. *Journal of Hydrology, 458–459,* 1–8.

Lal, R. (2001). Soil degradation by erosion. *Land Degradation and Development, 12,* 519–539.

Langland, M. J., Cronin, T. M., & US, G. S. (2003). *A summary report of sediment processes in Chesapeake Bay and watershed*. US Department of the Interior, US Geological Survey.

Legates, D. R., & McCabe, G. J., Jr. (1999). Evaluating the use of "goodness-of-fit" measures in hydrologic and hydroclimatic model validation. *Water Resources Research, 35,* 233–241.

Li, S., Li, J., & Zhang, Q. (2011). Water quality assessment in the rivers along the water conveyance system of the Middle Route of the South to North Water Transfer Project (China) using multivariate statistical techniques and receptor modeling. *Journal of Hazardous Materials, 195,* 306–317.

Li, S., Liu, W., Gu, S., Cheng, X., Xu, Z., & Zhang, Q. (2009). Spatio-temporal dynamics of nutrients in the upper Han River basin, China. *Journal of Hazardous Materials, 162,* 1340–1346.

Liew, M. W. V., Feng, S., & Pathak, T. B. (2012). Climate change impacts on streamflow, water quality, and best management practices for the Shell and Logan Creek Watersheds in Nebraska, USA. *International Journal of Agricultural and Biological Engineering, 5,* 13–34.

Luce, C. H., & Black, T. A. (1999). Sediment production from forest roads in western Oregon. *Water Resources Research, 35,* 2561–2570.

Luo, P., He, B., Takara, K., Razafindrabe, B. H. N., Nover, D., & Yamashiki, Y. (2011). Spatiotemporal trend analysis of recent river water quality conditions in Japan. *Journal of Environmental Monitoring, 13,* 2819–2829.

Ma, X., Li, Y., Zhang, M., Zheng, F., & Du, S. (2011). Assessment and analysis of non-point source nitrogen and phosphorus loads in the Three Gorges Reservoir Area of Hubei Province, China. *Science of the Total Environment, 412–413,* 154–161.

Mann, H. B. (1945). Nonparametric tests against trend. *Econometrica, 13,* 245–259.

McMahon, G., Alexander, R. B., & Qian, S. (2003). Support of total maximum daily load programs using spatially referenced regression models. *Journal of Water Resources Planning and Management, 129,* 315–329.

Meade, R. H., Dunne, T., Richey, J. E., Santos, U. D. M., & Salati, E. (1985). Storage and remobilization of suspended sediment in the lower Amazon River of Brazil. *Science, 228,* 488.

Mizugaki, S., Onda, Y., Fukuyama, T., Koga, S., Asai, H., & Hiramatsu, S. (2008). Estimation of suspended sediment sources using 137Cs and 210Pbex in unmanaged Japanese cypress plantation watersheds in southern Japan. *Hydrological Processes, 22,* 4519–4531.

Nearing, M. A., Jetten, V., Baffaut, C., Cerdan, O., Couturier, A., Hernandez, M., et al. (2005). Modeling response of soil erosion and runoff to changes in precipitation and cover. *Catena, 61,* 131–154.

Nelson, E. J., & Booth, D. B. (2002). Sediment sources in an urbanizing, mixed land-use watershed. *Journal of Hydrology, 264,* 51–68.

Nielsen, J. P. (1999). *Record extension and streamflow statistics for the Pleasant River, Maine*. US Department of the Interior, US Geological Survey.

Preston, S. D., & US, G. S. (2009). *SPARROW modeling: Enhancing understanding of the nation's water quality*. US Department of the Interior, US Geological Survey.

Rajaee, T., Nourani, V., Zounemat-Kermani, M., & Kisi, O. (2011). River suspended sediment load prediction: Application of ANN and wavelet conjunction model. *Journal of Hydrologic Engineering, 16,* 613.

Ran, Q., Su, D., Li, P., & He, Z. (2012). Experimental study of the impact of rainfall characteristics on runoff generation and soil erosion. *Journal of Hydrology, 424,* 99–111.

Reckhow, K. H. (1994). Water quality simulation modeling and uncertainty analysis for risk assessment and decision making. *Ecological Modelling, 72,* 1–20.

Runkel, R. L., Crawford, C. G., Cohn, T. A., & US, G. S. (2004). *Load estimator (LOADEST): A FORTRAN program for estimating constituent loads in streams and rivers*. US Department of the Interior, US Geological Survey.

Sakamoto, Y., Ishiguro, M., & Kitagawa, G. (1986). *Akaike information criterion statistics*. New York: Springer.

Schwarz, G. E. (2008). A Preliminary SPARROW Model of Suspended Sediment for the Conterminous United States. U.S. Geological Survey Open-File Report 2008–1205, 7 pp. Retrieved June 2019, from https://pubs.usgs.gov/of/2008/1205/ofr2008-1205.pdf.

Schwarz, G. E., Hoos, A. B., Alexander, R. B., & Smith, R. A. (2006). *The SPARROW surface water-quality model: Theory, application, and user documentation*. USGS Techniques and Methods Report. Book 6, Chapter B3. USGS, Washington, DC.

Shumway, R. H., Azari, R. S., & Kayhanian, M. (2002). Statistical approaches to estimating mean water quality concentrations with detection limits. *Environmental Science and Technology, 36,* 3345–3353.

Smith, R. A., Schwarz, G. E., & Alexander, R. B. (1997). Regional interpretation of water-quality monitoring data. *Water Resources Research, 33,* 2781–2798.

Smith, V. H. (1982). The nitrogen and phosphorus dependence of algal biomass in lakes: An empirical and theoretical analysis. *Limnology and Oceanography, 27,* 1101–1112.

Somura, H., Takeda, I., Arnold, J. G., Mori, Y., Jeong, J., Kannan, N., et al. (2012). Impact of suspended sediment and nutrient loading from land uses against water quality in the Hii River basin, Japan. *Journal of Hydrology, 450,* 25–35.

Sprague, L. A., & Lorenz, D. L. (2009). Regional nutrient trends in streams and rivers of the United States, 1993–2003. *Environmental Science and Technology, 43,* 3430–3435.

Trimble, S. W. (1997). Contribution of stream channel erosion to sediment yield from an urbanizing watershed. *Science, 278,* 1442–1444.

Tu, J. (2011). Spatial and temporal relationships between water quality and land use in northern Georgia, USA. *Journal of Integrative Environmental Sciences, 8,* 151–170.

Vogel, R. M., & Stedinger, J. R. (1985). Minimum variance streamflow record augmentation procedures. *Water Resources Research, 21,* 715–723.

Wang, Y., He, B., Duan, W., Li, W., Luo, P., & Razafindrabe, B. (2016). Source apportionment of annual water pollution loads in river basins by remote-sensed land cover classification. *Water, 8,* 361.

Webb, B. W., Phillips, J. M., Walling, D. E., Littlewood, I. G., Watts, C. D., & Leeks, G. (1997). Load estimation methodologies for British rivers and their relevance to the LOIS RACS (R) programme. *Science of the Total Environment, 194,* 379–389.

Woli, K. P., Nagumo, T., Kuramochi, K., & Hatano, R. (2004). Evaluating river water quality through land use analysis and N budget approaches in livestock farming areas. *Science of the Total Environment, 329,* 61–74.

Wolman, M. G. (1967). A cycle of sedimentation and erosion in urban river channels. *Geografiska Annaler. Series A. Physical Geography, 49,* 385–395.

Yue, S., Pilon, P., & Cavadias, G. (2002). Power of the Mann–Kendall and Spearman's rho tests for detecting monotonic trends in hydrological series. *Journal of Hydrology, 259,* 254–271.

Chapter 8
The Aral Sea Basin Crisis and Future Water–Climate–Food Nexus in Turkmenistan

8.1 Introduction

Water, land, and energy resources are critical to food security and ultimately to sustainable socio-economic development (Ellabban et al. 2014; Liu et al. 2018; Mohtar et al. 2019). Globally, water withdrawals and water use for food production account for 70 and 90% of total water withdrawals, respectively, (AQUASTAT 2016) and about 10% accounted for energy production and power generation in 2014 (IEA 2016). It is estimated that by 2050, global population growth will increase food demand by about 70%, and will increase by about three times by 2100 (Clay 2011; Food and Agriculture Organization 2009). The pressure associated with population growth will continue to be exacerbated by climate change, which will affect food production, hydropower, and many other important ways of water use (Conway et al. 2015). Therefore, to identify the linkages between water, energy, and food production can help us to fully understand and recognize the water–climate–food nexus for sustainable development under population growth and global warming (Dai et al. 2018; Mohtar and Daher 2015; Mortada et al. 2018).

In Central Asia, the vulnerability of water–climate–food security relationships is particularly acute (Jalilov et al. 2016) due to the arid or semi-arid climate (Bekturganov et al. 2016), transboundary rivers (Bernauer and Siegfried 2012), and rising temperature (Aleksandrova 2015; Farinotti et al. 2015) in this region. The Aral Sea Basin is the largest basin in Central Asia, which intersects all five Central Asian republics including Kazakhstan, Kyrgyzstan, Tajikistan, Turkmenistan, and Uzbekistan. With the rapid economic development and population explosion during recent years, the Aral Sea (the world's fourth largest lake in 1960) has already shrunk to almost one-tenth of its former size with the rise of toxic salt around the sea region, causing terrible ecological disasters, which was called the Aral Sea Basin crisis in the world (Micklin 2004; Spoor 1998).

The main reason is the significant decrease in the amount of water flowing from the Amu Darya River and Syr Darya River into the Aral Sea over the past 60 years, which directly caused a sharp decrease in the seawater level due to the vast amounts

W. Duan and K. Takara, *Impacts of Climate and Human Activities on Water Resources and Quality*, https://doi.org/10.1007/978-981-13-9394-5_8

of water for irrigation. Along with the rapid economic growth and population rise in this region, the future demand for water resources will certainly exacerbate the vulnerability of water–climate–food security relationships. For example, the main source of water in Turkmenistan is the Amu Darya River, a typical transboundary river between Turkmenistan and Uzbekistan, which originates from the Pamir and Tien Shan Mountains. Limited water resources, low irrigation efficiency, and a rapidly growing population and economy are driving the country to develop a sustainable resource management strategy (Bekchanov 2014; Bekchanov et al. 2016).

Therefore, this chapter firstly analyzes climate change impacts on water resources in the Aral Sea Basin, and then employs climate predictions to analyze the water balance for the Amu Darya River Basin and evaluate future water use, crop yields, land, and water productivities in Turkmenistan from 2016 to 2055. We also have explored the potential impact of future climate change on the relationship between water and food security in Turkmenistan.

8.2 Data and Methods

8.2.1 Study Area

The Aral Sea Basin is a transboundary river basin at the heart of the Eurasian continent (Fig. 8.1a). Its area is about 1.76 million km^2, which is shared by Tajikistan (about 99%), Turkmenistan (about 95%) and Uzbekistan (about 95%), Kyrgyzstan (about 59%), Kazakhstan (about 13%), northern Afghanistan (about 38%), and a very small part of Iran. The Aral Sea is mainly fed by the Syr Darya and Amu Darya rivers, and both rivers originate from Kopet Dag, western Tien Shan, and Pamirs.

The Amu Darya River is the largest river in Central Asia. It originates from the Pamiro-Alai Mountains and flows into the Aral Sea, with a total area of approximately 1.33×10^6 km^2 (Fig. 8.1b). The Vakhsh River and the Pyanj River are two tributaries. The Amu Darya River is mainly fed by rainfall and the melting of snow and ice, which is vulnerable to global warming. In recent years, rapid economic development and population growth in the basin are driving significant changes in land cover, water supply, and water demand patterns (Rakhmatullaev et al. 2017).

Turkmenistan lies in the lower and middle reaches of the Amu Darya River and is mainly covered (80%) by the Karakum Desert (Fig. 8.1b). The Amu Darya River is the main water source for all agricultural and non-agricultural uses in Turkmenistan, accounting for more than 80% of total freshwater resources. According to the water allocation agreements between all Central Asian countries (Protocol No. 566), Turkmenistan's share is about 35.8% of the total runoff from the Amu Darya River (Kostianoy et al. 2011). More than 90% of the water is consumed by irrigated agriculture (e.g., for cotton, wheat, rice, vegetables, fruits, and cucurbits) through an integrated water management system (e.g., the Karakum Canal), mainly in our focus provinces of Ahal, Dashoguz, Lebap, and Mary.

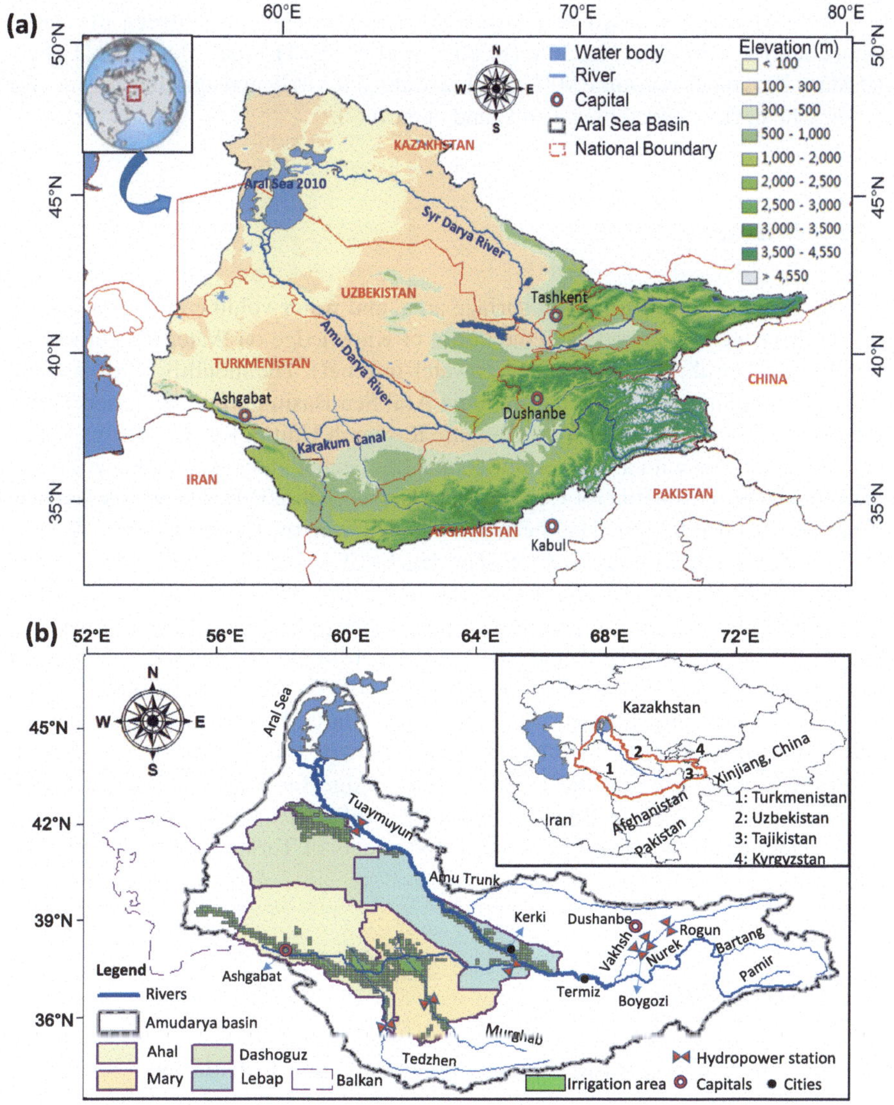

Fig. 8.1 Location of **a** the Aral Sea Basin, **b** the Amu Darya River Basin, river systems, and Turkmenistan

Turkmenistan is located in the middle and lower reaches of the Amu Darya River and is mainly covered by the Karakum Desert (80%) (Fig. 8.1b). The Amu Darya River is the main source of water for all agricultural and non-agricultural usage in Turkmenistan, which accounts for more than 80% of the total freshwater resources. According to the Water Distribution Agreement between all Central Asian countries (Protocol 566), Turkmenistan's share accounts for approximately 35.8% of the total flow of the Amu Valley based on the water allocation agreements between all Central Asian countries (Protocol No. 566) (Kostianoy et al. 2011). Ahal, Dashoguz, Lebap, and Mary Provinces consume 90% water resources for irrigated agriculture (e.g., for cotton, wheat, rice, vegetables, fruits, and cucurbits).

8.2.2 Datasets

The yearly water level (m), water surface area, and water volume in Aral Sea from 1911 to 2018 were obtained from the Portal of Knowledge on Water Resources and Ecology of Central Asia (http://www.cawater-info.net). The monthly, seasonal, and yearly precipitation and temperature in the Aral Sea Basin during the period 1901–2016 were computed based on the Climatic Research Unit (CRU TS v.4.01), which is a gridded dataset with a 0.5° resolution and has been widely used in the world and confirmed to be reasonable for Central Asia (Harris et al. 2014). The yearly land use during the period 1992–2015 has been downloaded from the European Space Agency (ESA) climate change initiative (CCI). The data has 37 categories with a 300 m spatial resolution and been confirmed to be reliable in Central Asia (Hollmann et al. 2013), which finally have been merged into 7 types (agricultural land, forest, grassland, water body, construction land, and bare land) according to the categories defined by the Intergovernmental Panel on Climate Change (IPCC) (Hurtt et al. 2009).

Most of the original datasets were from the Official Statistics of Turkmenistan (http://www.stat.gov.tm/), the World Bank Group (https://data.worldbank.org/country/turkmenistan), and the Portal of Knowledge for Water and Environmental Issues in Central Asia (http://www.cawater-info.net/). Other data came from the national or provincial energy strategy documents of Turkmenistan (Garabayeva 2012).

8.2.3 Methodology

8.2.3.1 The Aral Sea Basin Management Model and Calculation Process

The Aral Sea Basin Management Model (ASBmm), jointly developed by the Science Information Center of the Interstate Coordination Water Commission of Central Asia (SIC ICWC) and UNESCO-IHE, was used to compute water balance and determine water allocation in the Amu Darya River Basin (De Schutter 2008; Merks 2018). The aim is to develop a comprehensive assessment methodology for water sustainable development, taking into account socio-economic, environmental, energy, and climatic factors. ASBmm contains two models for the Amu Darya and Syr Darya watersheds. Each model contains four main modules, including a flow formation model, a water distribution model (WAm), a planning area model (PZm), and a socio-economic model. The water balance covers almost all hydrological processes, including evaporation, groundwater influx and outflow, seepage, return flow from the planning zones (PZs), PZs' extraction, and inflows and discharges into other parts (detailed information can be accessed from http://www.cawater-info.net/).

Figure 8.2 shows the hydrological scheme of the Amu Darya River Basin and the yellow quadrangles are the PZs for the four provinces in Turkmenistan. The PZm calculates the water balance of each water management political province under different climate, socio-economic, and water management scenarios related to river networks, including water balance in irrigation areas, backwater estimation (CDF), and calculations of agricultural output losses (in case of water shortages) on the basis of local resources (e.g., groundwater) and flow regulation by local reservoirs. PZm is used here to calculate water demand and to assess the land and water productivity of four interested provinces (Ahal, Dashoguz, Lebap, and Mary) in Turkmenistan.

Figure 8.3 shows the calculation framework for water requirements, water deficits, crop yield, agricultural output loss (in the absence of water), and land and water production in four provinces of Turkmenistan. In this study, the existing ASBmm model was used to determine the water balance and water distribution of future cross-border rivers (De Schutter 2008; Merks 2018). A regional climate model called REMO-0406 was to assess the possible impacts of climate changes on water resources. Three socio-economic scenarios, including the food security and diet change (FSD), export-oriented sustainable adaptation (ESA), and business as usual (BAU) scenarios, were adopted. BAU means "no change" in the future. The goal of FSD is to maximize food self-sufficiency based on population growth; ESA's goal is to achieve maximum food exports in the future with a food self-sufficiency rate of 80%. In all three cases, the demographic changes from 2016 to 2055 are based on demographic data for each planned area over the past decade. For the BAU scenario, future planting methods and yields are predicted based on trends in the last decade. The long-term target parameters adopted by the National Social and Economic Development Plan of Turkmenistan (2011–2030) are used to build future planting patterns and yields for FSD and ESA scenarios.

A coefficient of provision of water withdrawal limit from transboundary sources was selected to determine the water scenario. The coefficient is 50% in this study, suggesting that 50% of river discharges from the Amu Darya River could be taken during the growing season. An innovation scenario was also adopted, which means that development of innovative and efficient irrigation methods was used in irrigation. The main indicators evaluating the irrigation system contain water delivery, crop yields (cotton, wheat, and rice), and water productivity. Under the above circumstances, Turkmenistan's future water requirements and water deficits for irrigated land, agricultural yields of several major crops, land productivity (total output per hectare of irrigated area), and water productivity (gross production per m^3 water intake) were computed based on the following equations:

(1) The monthly standard net crop water requirements in planning zones are described by Eq. 8.1:

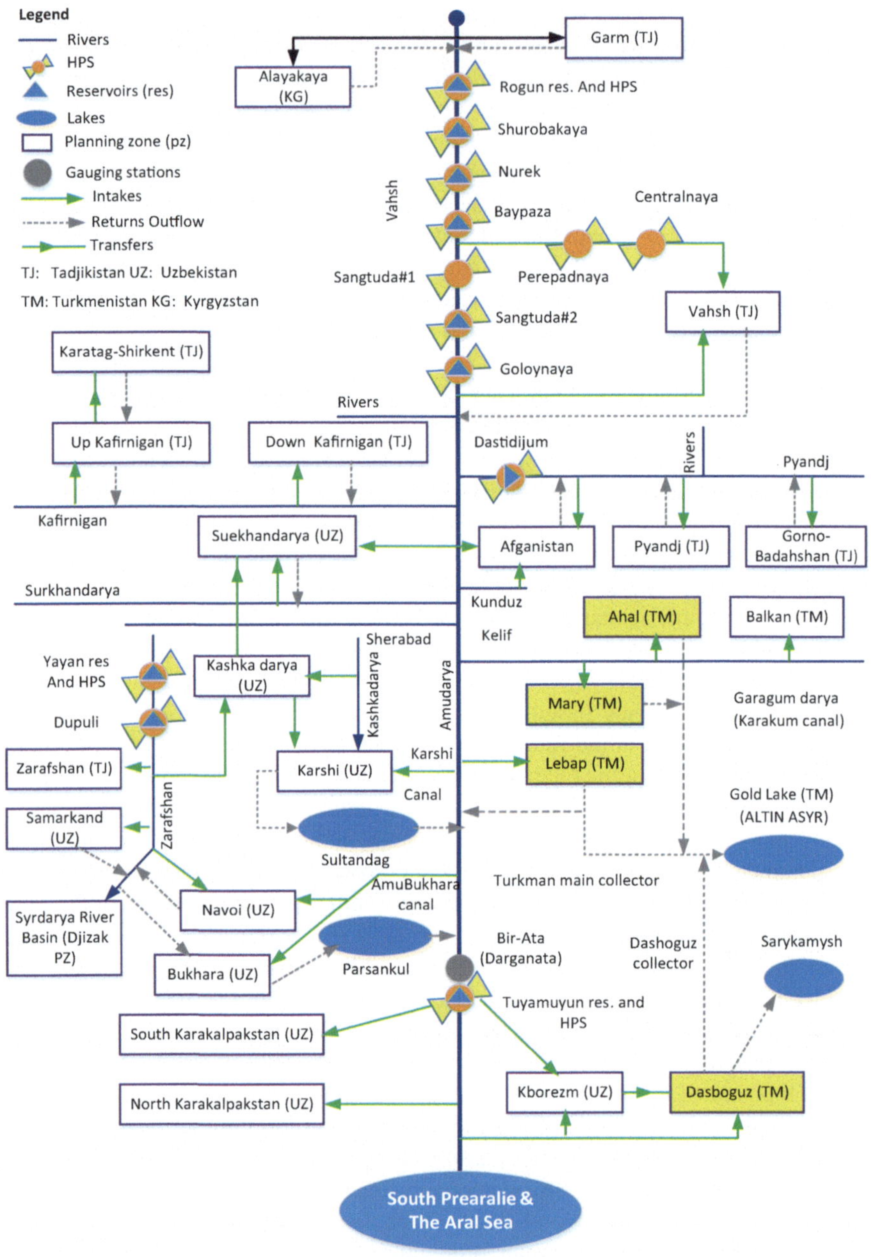

Fig. 8.2 Hydrological scheme of the Amu Darya River Basin and planning zones for four provinces (yellow background) in Turkmenistan (Modified from http://www.cawater-info.net/)

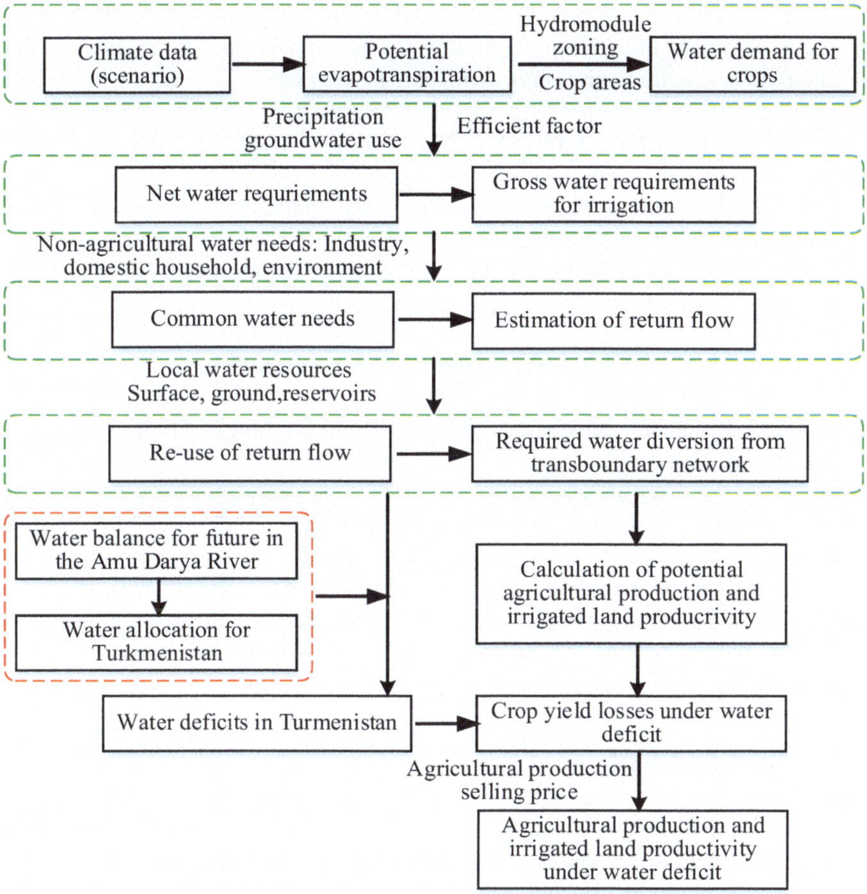

Fig. 8.3 Procedure for calculating water requirements and irrigated agriculture productivity

$$W_{zirn}(dt) = H_z * \sum wzm_r(dt) * Hz_r \tag{8.1}$$

where H_z is the total irrigated area of the four provinces (10^3 ha), $wzm_r(dt)$ is the average annual standard of irrigation for crop r (m^3/ha), and Hz_r is fraction of area under crop r in each province out of the total irrigated area of the four provinces. The predicted standard irrigated agricultural water demand, including losses in the irrigation network, is estimated as follows:

$$W_{zir}(dt) = W_{zirn}(dt) * \frac{ET(dt) - pzef(dt)}{ET_m(dt) - pzmef(dt)} \tag{8.2}$$

where $W_{zirn}(dt)$ represents the monthly standard net crop water requirements in PZ (10^6 m^3), $ET(dt)$ is the predicted reference evapotranspiration, $ET_m(dt)$ is the average annual reference evapotranspiration, and $pzmef(dt)$ is the average annual monthly effective precipitation for each province. The average annual reference evapotranspiration is estimated as follows:

$$\mathrm{ET}_m(dt) = 0.00144 * (25 + T(dt))^2 * (100 - L_m(dt)) \tag{8.3}$$

where $T(dt)$ is the predicted average monthly temperature (K), and $L_m(dt)$ is the average monthly humidity (%).

(2) Output losses calculate the average crop yield losses in case of water shortage for crop r for each province ($\delta \mathrm{Yz}_{gr}$, %) according to

$$\delta \mathrm{Yz}_{gr} = \mathrm{s}_r * (100 - \alpha) * \eta * wnz_r / \mathrm{ET}_r \tag{8.4}$$

where wnz_r is the necessary irrigation with an account of climate change in future, calculated as follows:

$$\alpha = 100 * Wlz / Wzdir \tag{8.5}$$

where Wlz is the irrigation water supply limit for a season (10^6 m^3), $Wzdir$ is the irrigated agriculture water demand, including losses in the irrigation network (within the boundaries of planning zones) over a season (10^6 m^3).

8.2.3.2 Statistical Analysis

The Mann–Kendall test was employed to compute trends of climate factors (temperature, precipitation), water requirements, food production, and socio-economic impacts from land production and water production under different scenarios during the study period. Given an independent and random dataset consisting of T values with the sample size n, the Mann–Kendall S Statistic is calculated as follows (Kendall 1975; Mann 1945):

$$S = \sum_{i=1}^{n-1} \sum_{j=i+1}^{n} \mathrm{sign}(T_j - T_i) \tag{8.6}$$

$$\mathrm{sign}(T_j - T_i) = \begin{cases} 1 & \text{if } T_j - T_i > 0 \\ 0 & \text{if } T_j - T_i = 0 \\ -1 & \text{if } T_j - T_i < 0 \end{cases} \tag{8.7}$$

where n is the sample size, T_j and T_i represent the variability of the above indicators at times j and i ($j > i$). The sample size (n) is 40. When $n \geq 10$, S is approximately normally distributed with mean $E(S) = 0$ and variance (σ^2), which can be computed as follows:

$$\sigma^2 = \frac{n(n-1)(2n+5) - \sum t_i(i)(i-1)(2i+5)}{18} \quad (8.8)$$

where $E(S)$ denotes the mean value of the S statistic, t_i is the number of ties to extent i. The summation term in the numerator is only used if the above data series contains tied values. The standard test statistic Z_S is calculated as follows:

$$Z = \begin{cases} \frac{S-1}{\sigma} & \text{for } S > 0 \\ 0 & \text{for } S = 0 \\ \frac{S+1}{\sigma} & \text{for } S < 0 \end{cases} \quad (8.9)$$

The test statistic Z_S is used as a measure of significance of trend of the above data series (e.g., see Duan et al. 2017 and Yue et al. 2002). When $|Z|$ is greater than $Z_{\alpha/2}$, where α represents the chosen significance level (e.g., 5% with $Z_{0.025} = 1.96$), the null hypothesis is invalid implying that the trend of the above data series is significant.

8.3 Results

8.3.1 Changes of Water Resources in the Aral Sea Basin

8.3.1.1 Changes of Water Level, Water Surface Area, and Water Volume

Figure 8.4 shows time series of water level (m), water surface area (103 m^2), and water volume (km^3) in Aral Sea from 1911 to 2018, suggesting that a decreasing

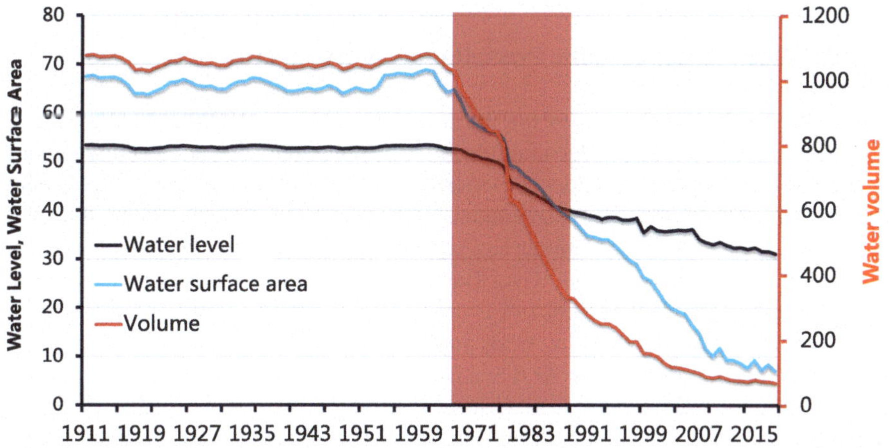

Fig. 8.4 Time series of water level (m), water surface area (10^3 m^2), and water volume (km^3) in Aral Sea from 1911 to 2018

Fig. 8.5 Changes of lake surface area in Aral Sea from 1973 to 2018

trend for all three indices. Concretely, the water surface area decreased from 67.5 × 10^3 m^2 in 1911 to 6.99 × 10^3 m^2 in 2018, the water volume decreased from 1078 km^3 in 1911 to 69.31 km^3 in 2018, and the water level decreased from 53.32 m in 1911 to 31.03 m in 2018. Figure 8.5 illustrates changes in lake surface area in Aral Sea from 1973 to 2018, which indicates that the Aral Sea has already shrunk to almost one-tenth its former size. In 1987–1988, the Aral Sea split into two parts, and the Large and Small Aral Sea Basins were created. In 2018, the surface area was only 6.99 × 10^3 m^2.

8.3.1.2 Trend Analysis of the Precipitation and Temperature

Figure 8.8 shows distributions of the mean annual values in the precipitation (mm), temperature (°C), and linear decadal trends in the annual precipitation (mm/decade) and annual temperature (°C/decade) during 1901–2015 in the Aral Sea Basin. Figure 8.8a indicates that the mean annual precipitation ranged from 100 to 700 mm, which was higher in the mountainous regions compared with the other regions. Except for the upper Syr Darya River Basin, the same distribution was detected in the precipitation trends (Fig. 8.8c).

Figure 8.8b indicates that the mean annual temperature was much higher in the middle and lower regions of the Aral Sea Basin compared with the high mountainous regions, ranging from around −5 °C to around 15 °C. As shown in Fig. 8.8d, an increasing trend with the significance on a 95% significance level was found for all grid cells of the basin, ranging from approximately 0.04 °C/decade to 0.18 °C/decade. The northwestern region of the basin showed a larger increase compared with the

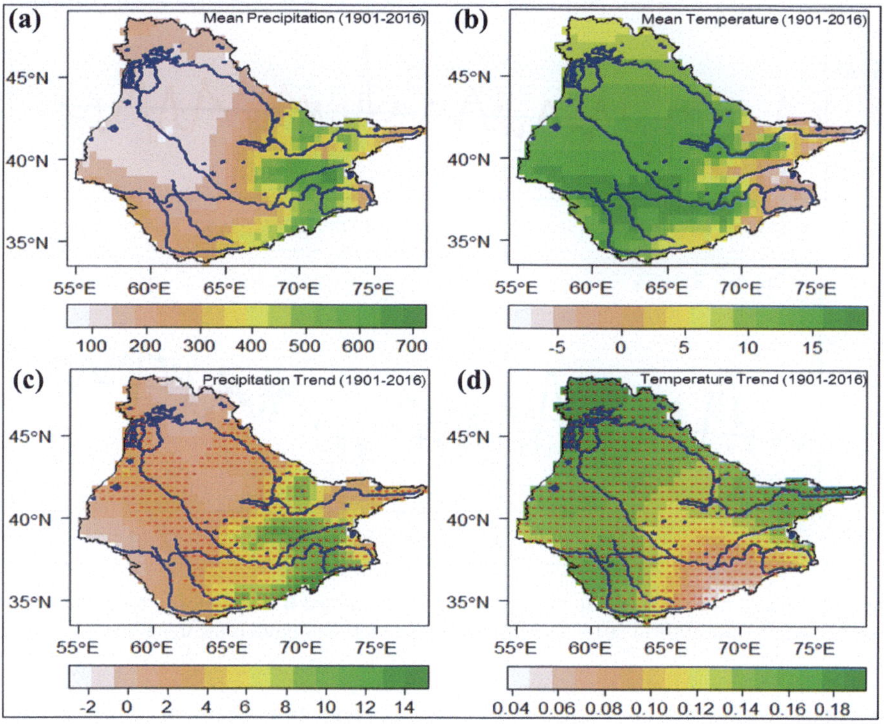

Fig. 8.6 Maps of the mean annual values in the **a** precipitation (mm), **b** temperature (°C) and linear decadal trends in the **c** precipitation (mm/decade), and **d** temperature (°C/decade) during 1901–2016 in the Aral River Basin. Red pluses indicate the grid points with changes that are significant on a 95% significance level

other regions, up to 0.18 °C/decade, suggesting more sensitive to global warming in the plain, which has accelerated the glacier melting in the mountains regions (Zou et al. 2019) (Fig. 8.6).

Figure 8.7 shows time series and trends in precipitation (mm) and temperature (°C) in the whole Aral Sea Basin from 1901 to 2016, which reveals that a significant upward trend was detected for both precipitation and temperature. This kind of trends was also predicted to occur in the future, probably causing pressure on water supply and food production in the Aral Sea Basin.

8.3.1.3 Trend Analysis of the Precipitation and Temperature

Figure 8.8 shows Time series of the water delivery to the Aral Sea from the Amu Darya River Delta and the Syr Darya River Delta for both growing season and non-growing season from 1992 to 2018, which suggests that the water delivery to the Aral Sea and the Amudarya river Delta decreased significantly for both growing season

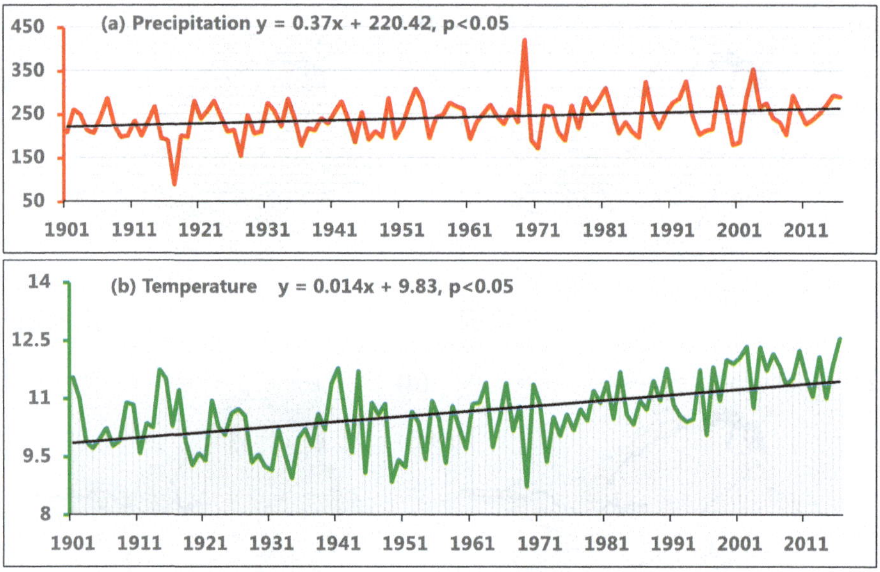

Fig. 8.7 Time series and trends in the **a** precipitation (mm) and **b** temperature (°C) in the whole Aral Sea Basin from 1901 to 2016. The straight line in each sub-figure is the trend line

and non-growing season. Figure 8.9 shows the changes in the main types of land use in the Aral Sea Basin from 1992 to 2015. Generally, grassland is the largest land cover, followed by agricultural land and bare land. A noticeable increase on a 95% statistical significance level was found in the agricultural land and construction land exhibited during 1992–2015, implying that the rapid expansion of the agricultural land consumed more water resources.

8.3.2 Future River Runoff in Amu Darya River

Figure 8.10 shows that the annual runoff exhibits a slight increase from 2016 to 2055, with the Z value at 0.90 (not significant). The year of 2026 has the highest annual runoff, up to 104×10^9 m^3/year. The average annual runoff during the period of 2016–2055 is about 66.77×10^9 m^3/year, which is 1.68×10^9 m^3/year less than that for the period 1933–2015. The average annual runoff for Pyandj, Vakhsh, Kafirnigan, Surkhandarya, and Kunduz Rivers during the period 2016–2055 is approximately 34.50×10^9, 19.12×10^9, 5.34×10^9, 3.42×10^9, and 4.39×10^9 m^3/year, respectively. Under the REMO 0406 scenario, the Pyandj River will have the largest decrease in the future, up to 0.86×10^9 m^3/year. In addition, although the annual runoff from 2016 to 2055 is small, it varies greatly from month to month. The river runoff is expected to decrease from June to August, and increase in the months of April and the growing season.

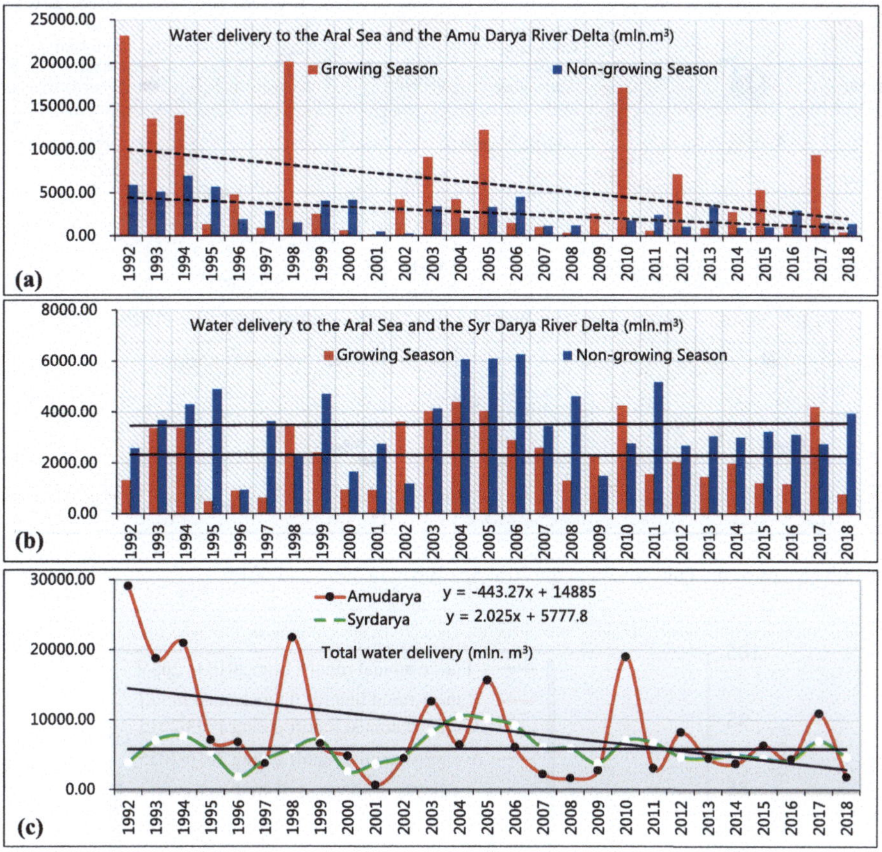

Fig. 8.8 Time series of the water delivery to the Aral Sea from the Amu Darya River Delta and the Syr Darya River Delta for both growing season and non-growing season from 1992 to 2018

8.3.3 Future Water–Climate–Food Nexus in Turkmenistan

8.3.3.1 Water Requirements and Deficits During Growing Season

The time series of water requirements and water deficits for the growing season (April–September) in Turkmenistan from 2016 to 2055 are shown in Fig. 8.11a, b. The water requirement under the BAU scenario is clearly less than that in both FSD and ESA scenarios, and the difference between them continues to increases, up to 4.40×10^9 m^3/month by 2055. Similarly, the water deficits under the BAU scenario are clearly less than that in both FSD and ESA scenarios. Under the ESA, FSD, and BAU scenarios, the estimated water shortages are 7.88×10^9 m^3, 7.47×10^9 m^3, and 4.33×10^9 m^3 in 2055, respectively.

Table 8.1 shows the results of the Mann–Kendall test (Z value) of water requirements and water deficits for growing season under three scenarios (BAU, FSD, and

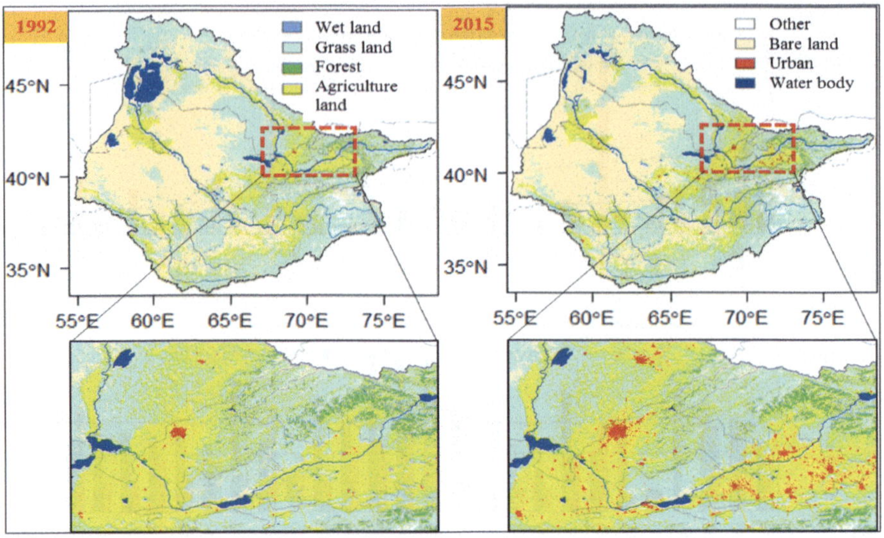

Fig. 8.9 Maps of the land covers in the Aral Sea Basin in 1992 and 2015

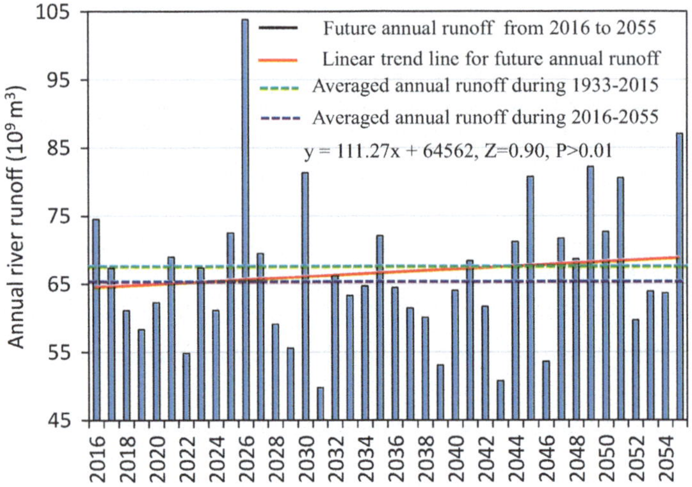

Fig. 8.10 Annual runoff of the Amu Darya River under the REMO 0406 scenario from 2016 to 2055. The red line is the linear trend line, and the dashed lines are the average annual river runoff for history (1933–2015) and future (2016–2055)

ESA) from 2016 to 2055. With the exception of Lebap and Mary Provinces, water requirements in all other provinces have shown a downward trend. Among them, from 2016 to 2055, the BAU scenario had a significant decreasing trend at the 95% confidence in all four provinces. Water requirements in Lebap Province under the ESA

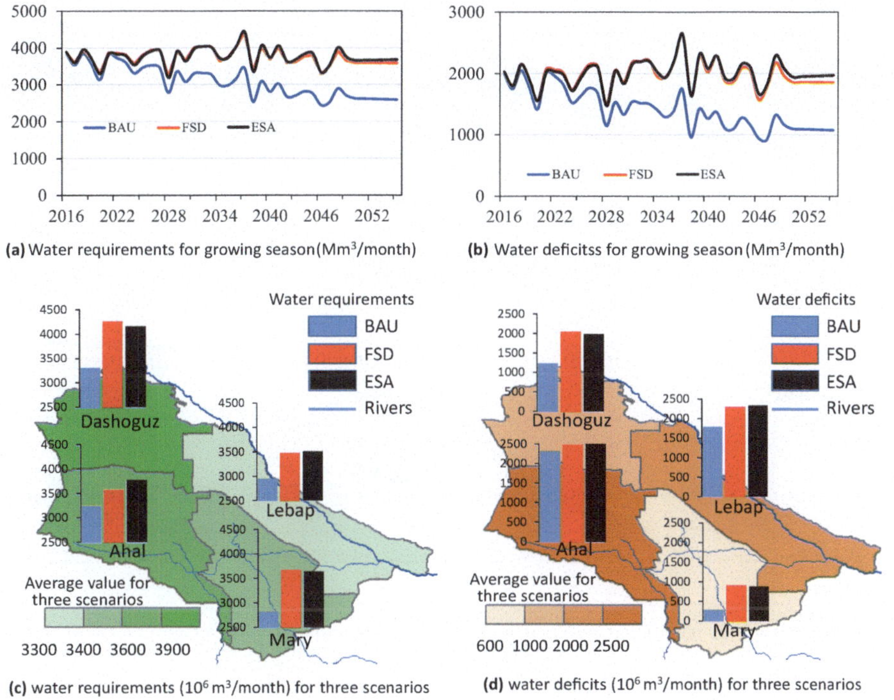

Fig. 8.11 Time series of **a** water requirements (10^9 m^3/month) and **b** water deficits (10^9 m^3/month) for growing season under three scenarios (BAU, FSD, and ESA) from 2016 to 2055. The annual mean **c** water requirements (10^6 m^3/month) and **d** water deficits (10^6 m^3/month) during growing season under BAU, FSD, and ESA scenarios in Turkmenistan and its four provinces from 2016 to 2055

Table 8.1 Mann–Kendall test (Z value) of water requirements and water deficits for growing season under three scenarios (BAU, FSD, and ESA) from 2016 to 2055

Regions	Water requirements			Water deficits		
	BAU	FSD	ESA	BAU	FSD	ESA
Ahal	−6.28	−3.88	−1.50	−5.98	−3.53	−0.94
Dashoguz	−5.25	0.03	−0.78	−4.60	0.90	0.55
Lebap	−5.98	0.62	2.67	−5.98	0.64	2.85
Mary	−6.63	0.15	0.38	−5.44	2.16	2.85
Turkmenistan	−6.56	−1.85	−0.64	−6.28	−0.92	0.76

scenario shows a significant increasing trend, indicating that water deficits during the growing season would increase (95% confidence). As with water requirements, water deficits exhibit a significant decreasing trend under the BAU scenario from 2016 to 2055 in Turkmenistan. A significant increasing trend is also found for water deficits under both the FSD and ESA scenarios in Mary Province.

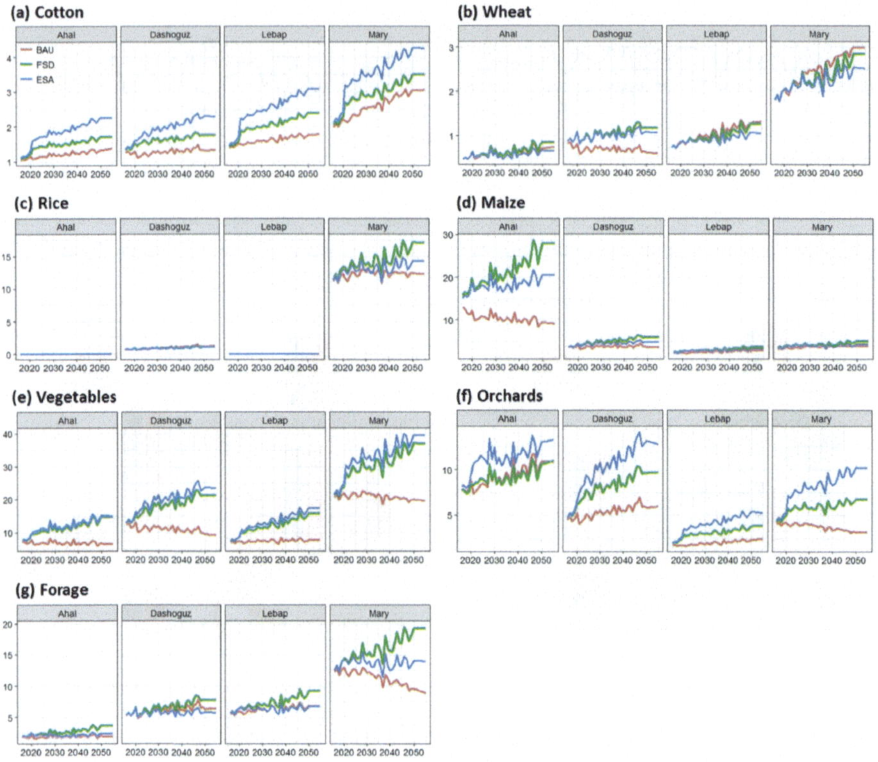

Fig. 8.12 Crop yield (t/ha) of **a** cotton, **b** wheat, **c** rice, **d** maize, **e** vegetables, **f** orchards, and **g** forage under BAU, FSD, ESA scenarios in four provinces of Turkmenistan from 2016 to 2055

Figure 8.11c, d show the average annual water requirements and annual water deficits in the growing seasons of the three scenarios from 2016 to 2055. As shown in the figure, the estimated average water demand in these three scenarios is the largest in Dashaguz, with a value of 3.90×10^9 m^3, followed by Ahal, Mary, and Leap. In addition to Ahal Province, FSD scenario in other provinces have the largest water requirements, followed by the ESA and BAU scenarios. In terms of water scarcity, Dashaguz Province has the highest average value of 2.50×10^9 m^3 in these three scenarios, while Mary Province has dropped to 0.60×10^9 m^3. This may be because there are less irrigated areas and local water resources in Mary Province. As with water requirements, the FSD scenario shows that all provinces have the largest water deficits, especially in the Ahal (2.50×10^9 m^3).

8.3.3.2 Agricultural Productivity

Unsurprisingly, the agricultural productivity results shown in Fig. 8.12 and Tables 8.2 and 8.3 show considerable time variations. Under the FSD and ESA scenarios, production of seven major crops in all four provinces is expected to increase in 2016–2055, including cotton, wheat, rice, corn, vegetables, orchards, and pastures, most of which are significant at the 95% confidence level. For example, by 2055, cotton yields are expected to increase to 1.71 t/ha, 1.76 t/ha, 2.39 t/ha, 3.51 t/ha under the

Table 8.2 Crop yields in the four provinces in 2030 and 2055 under different scenarios

Province	Scenario	BAU		FSD		ESA	
	Year	2030	2055	2030	2055	2030	2055
Ahal	Cotton	1.24	1.38	1.53	1.71	1.88	2.25
	Wheat	0.7	0.75	0.71	0.86	0.65	0.67
	Rice	0	0	0	0	0	0
	Maize	11.65	9.19	22.23	28.06	19.69	20.58
	Vegetables	7.73	6.62	12.06	14.68	12.83	15
	Orchards	9.85	10.87	9.9	10.96	12.41	13.27
	Forage	2.16	2	2.88	3.66	2.43	2.37
Dashoguz	Cotton	1.31	1.33	1.68	1.76	2	2.29
	Wheat	0.77	0.61	1.13	1.19	1.11	1.08
	Rice	1.03	1.3	0.97	1.11	0.95	1.06
	Maize	4.04	3.66	5.33	6.01	4.93	4.83
	Vegetables	12.13	9.37	19.16	21.48	21.38	23.61
	Orchards	5.53	5.94	8.55	9.75	10.8	12.76
	Forage	6.58	6.49	7.03	7.86	6.32	5.78
Lebap	Cotton	1.58	1.78	1.99	2.39	2.44	3.08
	Wheat	0.98	1.33	0.93	1.27	0.88	1.06
	Rice	0	0	0	0	0	0
	Maize	2.49	2.92	2.99	3.84	2.87	3.3
	Vegetables	7.4	7.72	11.43	15.96	12.46	17.52
	Orchards	1.88	2.43	2.98	3.89	3.94	5.24
	Forage	6.13	6.81	6.96	9.27	6.27	6.71
Mary	Cotton	2.5	3.06	2.86	3.51	3.45	4.26
	Wheat	2.41	3	2.18	2.86	2.14	2.53
	Rice	12.61	12.32	13.66	17.14	12.83	14.24
	Maize	3.78	3.97	4.17	5.16	4.01	4.36
	Vegetables	21.78	19.71	29.32	37.07	32.23	39.62
	Orchards	3.98	3.18	5.76	6.82	7.91	10.18
	Forage	12.1	8.92	14.99	19.3	13.57	13.99

Table 8.3 Results of the Mann–Kendall test (Z value) for six types of corps in Ahal, Dashoguz, Lebap, and Mary from 2016 to 2055

Provinces	Scenarios	Cotton	Wheat	Rice	Maize	Vegetables	Orchards	Forage
Ahal	BAU	7.19	5.50		−4.96	−1.79	6.14	2.20
	FSD	7.38	6.04		6.82	6.86	5.74	6.60
	ESA	8.12	3.93		4.73	6.78	5.55	2.43
Dashoguz	BAU	4.46	−3.50	7.13	0.77	−4.77	5.45	4.07
	FSD	6.54	5.12	6.40	6.98	7.03	7.26	6.39
	ESA	7.62	3.76	6.21	4.33	6.56	7.33	1.77
Lebap	BAU	7.33	7.24		6.35	3.09	6.96	5.11
	FSD	7.94	6.70		6.93	7.82	7.12	6.85
	ESA	8.25	5.45		5.61	7.80	7.43	2.48
Mary	BAU	7.98	8.00	0.56	6.43	−4.18	−7.02	−6.75
	FSD	7.49	6.75	6.00	6.31	6.64	5.77	6.51
	ESA	7.65	5.57	3.64	3.77	7.05	7.11	0.43

FED scenario and to 13.27 t/ha, 23.61 t/ha, 5.24 t/ha, 10.18 t/ha under the ESA scenario in Ahal, Dashoguz, Lebap, and Mary Provinces, respectively. Wheat yields are predicted to increase to 10.96 t/ha, 9.75 t/ha, 3.89 t/ha, 6.82 t/ha under the FED scenario, and to 13.27 t/ha, 23.61 t/ha, 5.24 t/ha, 10.18 t/ha under the ESA scenario in Ahal, Dashoguz, Lebap, and Mary Provinces, respectively. These estimated productivities BAU scenario are less than the yields under the FED and ESA scenarios in 2055. We can also find that the changes in yield between different years and between different provinces seem to be large, mainly due to the difference of various climatic factors in different regions.

Figure 8.12 and Tables 8.2 and 8.3 also clearly show that under these three scenarios, most of the crops in the province of Mary (including cotton, wheat, rice, vegetables, and forage) have higher yields, while the lower yields of wheat, rice, and forage in Ahal Province. Among them, the cotton yield in the province of Mary is about twice that of the Akhal Province, and the wheat and pasture are about three and five times, respectively. The dramatic increase in forage may reflect a significant increase in livestock in this province. Almost all of Turkmenistan's rice is produced in the province of Mary (Fig. 8.12), and by 2055, it is estimated to increase to 12.32, 17.14, and 14.24 t/ha under the BAU, FSD, and ESA scenarios, respectively. Under the BAU, FSD, and ESA scenarios, maize productivity in Ahal Province is the highest, reaching 9.19, 28.06, and 20.58 t/ha in 2055, respectively.

8.3.3.3 Land and Water Productivity

The estimated annual revenue of irrigated agriculture (Millions of USD), revenue losses in irrigated agriculture (USD), land productivity (USD/ha), and irrigation

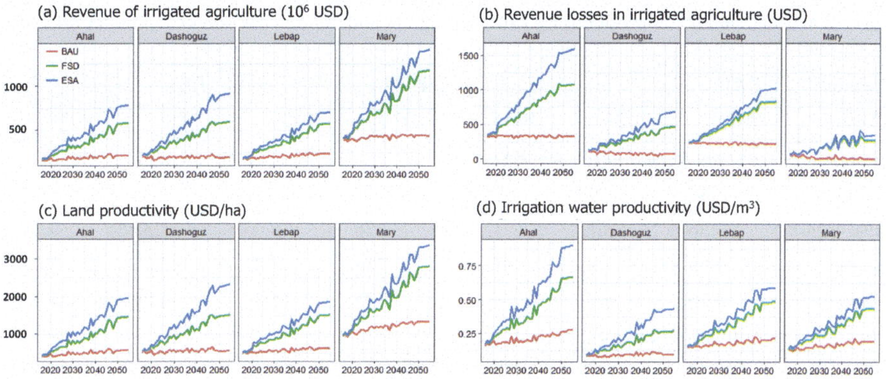

Fig. 8.13 Time series of **a** revenue of irrigated agriculture (10^6 USD), **b** revenue losses in irrigated agriculture (USD), **c** land productivity (USD/ha), and **d** irrigation water productivity (USD/m^3) from 2016 to 2055

water productivity (USD/m^3) are shown in Fig. 8.13 and Table 8.4, which suggests that a significant increasing trend for all four provinces under the FSD and ESA scenarios from 2016 to 2055. Under the FSD and ESA scenarios in Mary Province, the estimated annual revenue from irrigated agriculture is predicted to increase from 6.55×10^8 USD in 2015 to 11.80×10^8 USD and 14.20×10^8 USD in 2055, respectively, and the estimated annual land productivity is also predicted to increase from 1125.28 USD/ha in 2015 to 2813.67 USD/ha and 3373.65 USD/ha in 2055, respectively. Under the FSD and ESA scenarios in Ahal Province, the annual revenue losses in irrigated agriculture is predicted to increase from 283.54 USD in 2015 to 1083.19 USD and 1595.17 USD in 2055, respectively, and the estimated annual irrigation water productivity is also predicted to increase from 704.91 USD/ha in 2015 to 1469.6 USD/ha and 1968.02 USD/ha in 2055.

As shown in Fig. 8.13 and Table 8.4, BAU scenario is predicted to have the lowest value for all land and water productivity indicators. Under the BAU scenario, from 2016 to 2055, the estimated annual revenue losses of irrigated agriculture in Mary Province exhibits a downward trend. On the contrary, the largest value for all land and water productivity indictors is found in the ESA scenario, which indicates that under Turkmenistan's export-oriented sustainable adaptation, the income and losses of irrigated agriculture are likely to reach the highest level.

8.4 Discussions

The above results indicate that a sharp decrease in the amount of water flowing from the Amu Darya River and Syr Darya River into the Aral Sea over the past 60 years directly caused the Aral Sea Basin crisis. The Aral Sea and its fringing wetlands were an important resource for the agriculture, animal husbandry, fur

Table 8.4 Land and water productivity s in 2015, 2020, and 2055 in Turkmenistan under different scenarios

Scenario	BAU			FSD		ESA	
Year	2015	2020	2055	2020	2055	2020	2055
Revenue of irrigated agriculture (10^8 USD)							
Ahal	3.87	1.94	2.24	2.22	5.88	2.34	7.87
Dashoguz	4.36	2.12	1.98	2.76	6.03	2.85	9.21
Lebap	3.99	1.93	2.44	2.14	5.81	2.21	7.06
Mary	6.55	4.11	4.45	4.73	11.8	4.86	14.2
Revenue losses in irrigated agriculture (USD)							
Ahal	283.54	316.3	335.57	362.34	1083.19	378.55	1595.17
Dashoguz	40.45	79.11	80.51	126.17	470.67	129.47	689.32
Lebap	32.68	234.83	232.17	267.98	836.53	272.88	1030.13
Mary	26.86	39.44	13.59	77.12	283.08	77.85	354.39
Land productivity (USD/ha)							
Ahal	704.91	487.5	584.65	546.33	1469.6	576.57	1968.02
Dashoguz	1110.49	578.81	561.34	692.57	1526.52	714.37	2331.11
Lebap	1114.25	531.52	636.62	585.56	1521.27	604.37	1864.72
Mary	1125.28	1066.04	1344.02	1113.72	2813.67	1145.45	3373.65
Irrigation water productivity (USD/m^3)							
Ahal	0.23	0.2	0.28	0.23	0.67	0.25	0.9
Dashoguz	0.12	0.1	0.1	0.13	0.28	0.14	0.44
Lebap	0.18	0.16	0.22	0.18	0.49	0.19	0.59
Mary	0.19	0.15	0.2	0.17	0.44	0.17	0.53

trapping, and fishing industries during early Soviet Union times (Micklin 2004); however, since the 1950s, the irrigated land (especially for cotton) have increased dramatically, which have significantly affected the water supply and water demand patterns. During the late 1960s, the amount of water evaporating from the Aral Sea become greater than the amount of water flowing from the Amu Darya River and Syr Darya River into the lake, so lake levels significantly declined in the 1970s and 1980s (see Fig. 8.4). Climate change may exacerbate the challenge of meeting the needs of Turkmenistan's water–climate–food security nexus. Firstly, increasing temperature has been proven to increase the rate of evaporation and change crop phenology (Chmielewski et al. 2004; Peã Uelas et al. 2009), such as to shorten the growing season for cotton and therefore reduce water requirements (Wang et al. 2008). In addition, monthly precipitation will vary dramatically in the future, which will bring challenges for freshwater supply and water resources management.

Technological innovation can mitigate or even eliminate climate risks in water–climate–food security (Smithers and Blay-Palmer 2001), which should be placed in a priority position in the Amu Darya River Basin. For example, due to the low level of technical capacity, irrigation systems in Central Asia have wasted large amounts of water resources during crop irrigation and resulted in loss or degradation of fertile land, river degradation, and other environmental consequences (Zonn et al. 2018). In this chapter, therefore, an innovative solution was simulated to improve water use efficiency, which can offset the negative impacts of climate change.

Finally, limitations and uncertainty analysis are necessary. Clearly, climate models have a degree of uncertainty about future precipitation and temperature predictions, which may affect our ability to assess crop water requirements during the growing season (Knutti and Sedláček 2013). Secondly, it is difficult to predict the future runoff since the Amu Darya River is a river that is mainly supplied by ice and snow melting (Sorg et al. 2012). Thirdly, incomplete historical datasets, especially historical datasets from Afghanistan, introduce yet another uncertainty into the modeling via ASBmm.

8.5 Conclusions

In this chapter, climate change impacts on water resources in the Aral Sea Basin were analyzed, and future water–climate–food security nexus were estimated for four provinces of Turkmenistan from 2016 to 2055. Results could be summarized as follows: (1) A sharp decrease in the amount of water flowing from the Amu Darya River and Syr Darya River into the Aral Sea over the past 60 years directly caused the Aral Sea Basin crisis; (2) The expansion of the construction land and agricultural land increased the water consumption, exacerbating the stress of the water resources; (3) Changes in precipitation and temperatures are predicted to exacerbate the challenges of meeting needs of the water–climate–food security nexus in Turkmenistan; (4) Under the BAU, FSD, and ESA scenarios, a decreasing trend was found in water requirements and water deficits during the growing seasons from 2016 to 2055 in most provinces; (5) Under all three scenarios, production of seven major crops including cotton, wheat, rice, maize, vegetables, orchards, and forage, are predicted to exhibit a upward trend for four provinces for the period of 2016–2055; and (6) The highest mean income (about 7×10^8 USD/year) and losses of irrigated agriculture (about 1.50×10^3 USD/year) were found in Mary Province under all three scenarios, and the largest mean land (about 800 USD/ha/year) and irrigation water productivities (about 0.40 USD/m^3/year) were predicted in Ahal Province.

References

Aleksandrova, M. (2015). Water scarcity under climate change: Impacts, vulnerability and risk reduction in the agricultural regions of Central Asia. Ph.D. thesis, Ca'Foscari University of Venice.

AQUASTAT. (2016). FAO's information system on water and agriculture. Retrieved July 2019, from https://www.fao.org.

Bekchanov, M., Ringler, C., Bhaduri, A., & Jeuland, M. (2016). Optimizing irrigation efficiency improvements in the Aral Sea Basin. *Water Resources and Economics, 13,* 30–45.

Bekchanov, M. B. (2014). Efficient water allocation and water conservation policy modeling in the Aral Sea Basin. Ph.D. thesis, Rheinischen Friedrich-Wilhelms-Universität Bonn.

Bekturganov, Z., Tussupova, K., Berndtsson, R., Sharapatova, N., Aryngazin, K., & Zhanasova, M. (2016). Water related health problems in Central Asia-a review. *Water, 8,* 219.

Bernauer, T., & Siegfried, T. (2012). Climate change and international water conflict in Central Asia. *Journal of Peace Research, 49,* 227–239.

Chmielewski, F. M., Muller, A., & Bruns, E. (2004). Climate changes and trends in phenology of fruit trees and field crops in Germany, 1961–2000. *Agricultural and Forest Meteorology, 121,* 69–78.

Clay, J. (2011). Freeze the footprint of food. *Nature, 475,* 287–289.

Conway, D., Garderen, E. A. V., Deryng, D., Dorling, S., Krueger, T., Landman, W., et al. (2015). Climate and southern Africa's water-energy-food nexus. *LSE Research Online Documents on Economics* (vol. 5, pp. 837–846).

Dai, J., Wu, S., Han, G., Weinberg, J., Xie, X., Wu, X., et al. (2018). Water-energy nexus: A review of methods and tools for macro-assessment. *Applied Energy, 210,* 393–408.

De Schutter, J. (2008). Communicating the Issues of the Aral Sea Basin Long-Term Vision or the Aral Sea Basin the Aral Sea Basin management model (ASB-MM). In *Transboundary water resources: A foundation for regional stability in Central Asia* (pp. 29–42). Springer.

Duan, W., He, B., Sahu, N., Luo, P., Nover, D., Hu, M., et al. (2017). Spatiotemporal variability of Hokkaido's seasonal precipitation in recent decades and connection to water vapour flux. *International Journal of Climatology, 37,* 3660–3673.

Ellabban, O., Abu-Rub, H., & Blaabjerg, F. (2014). Renewable energy resources: Current status, future prospects and their enabling technology. *Renewable and Sustainable Energy Reviews, 39,* 748–764.

Farinotti, D., Longuevergne, L., Moholdt, G., Duethmann, D., Mölg, T., Bolch, T., et al. (2015). Substantial glacier mass loss in the Tien Shan over the past 50 years. *Nature Geoscience, 8,* 716.

Food and Agriculture Organization. (2009). Global agriculture towards 2050 (Rome, 2009). Retrieved August 2019, from http://www.fao.org/fileadmin/templates/wsfs/docs/Issues_papers/HLEF2050_Global_Agriculture.pdf.

Garabayeva, B. (2012). National program of Turkmenistan on socio-economic development for the period 2011–2030. Retrieve July 2019, from https://ims.undg.org/downloadFile/a303c0f7855ef9a8cceec3133577fd61b594acd58284b04098da77e65f63a6e3.

Harris, I., Jones, P. D., Osborn, T. J., & Lister, D. H. (2014). Updated high-resolution grids of monthly climatic observations-the CRU TS3. 10 Dataset. *International Journal of Climatology, 34,* 623–642.

Hollmann, R., Merchant, C. J., Saunders, R., Downy, C., Buchwitz, M., Cazenave, A., et al. (2013). The ESA climate change initiative: Satellite data records for essential climate variables. *Bulletin of the American Meteorological Society, 94,* 1541–1552.

Hurtt, G. C., Chini, L. P., Frolking, S., Betts, R. A., Feddema, J., Fischer, G., et al. (2009). Harmonisation of global land-use scenarios for the period 1500–2100 for IPCC-AR5.

IEA. (2016). World Energy Outlook 2016. Retrieve August 2019, from https://www.iea.org/publications/freepublications/publication/WorldEnergyOutlook2016ExecutiveSummaryEnglish.pdf.

Jalilov, S., Keskinen, M., Varis, O., Amer, S., & Ward, F. A. (2016). Managing the water–energy–food nexus: Gains and losses from new water development in Amu Darya River Basin. *Journal of Hydrology, 539,* 648–661.

Kendall, M. G. (1975). *Rank correlation methods: Charles Griffin*. London.

Knutti, R., & Sedláček, J. (2013). Robustness and uncertainties in the new CMIP5 climate model projections. *Nature Climate Change, 3,* 369–373.

Kostianoy, A. G., Lebedev, S. A., & Solovyov, D. M. (2011). Satellite monitoring of water resources in Turkmenistan. In *Fifteenth international water technology conference, IWTC-15, 31 March–02 April 2011,* Alexandria, Egypt.

Liu, J., Mao, G., Hoekstra, A. Y., Wang, H., Wang, J., Zheng, C., et al. (2018). Managing the energy-water-food nexus for sustainable development. *Applied Energy, 210,* 377–381.

Mann, H. B. (1945). Nonparametric tests against trend. *Econometrica, 13,* 245–259.

Merks, J. (2018). The Water efficiency paradox: The significance of spatial scale and knowledge exchange in irrigation water management. Master thesis, Delft University of Technology.

Micklin, P. (2004). *The Aral Sea crisis, dying and dead seas climatic versus anthropic causes* (pp. 99–123). Springer.

Mohtar, R. H., & Daher, B. (2015). Water–energy–food (WEF) Nexus Tool 2.0: Guiding integrative resource planning and decision-making. *Water International, 40,* 748–771.

Mohtar, R. H., Shafiezadeh, H., Blake, J., & Daher, B. (2019). Economic, social, and environmental evaluation of energy development in the Eagle Ford shale play. *Science of the Total Environment, 646,* 1601–1614.

Mortada, S., Najm, M. A., Yassine, A., Fadel, M. E., & Alamiddine, I. (2018). Towards sustainable water-food nexus: An optimization approach. *Journal of Cleaner Production, 178,* 408–418.

Peã Uelas, J., Rutishauser, T., & Filella, I. (2009). Ecology. Phenology feedbacks on climate change. *Science, 324,* 887–888.

Rakhmatullaev, S., Abdullaev, I., & Kazbekov, J. (2017). Water-energy-food-environmental nexus in Central Asia: From transition to transformation. In S. Zhiltsov, I. Zonn, A. Kostianoy, & A. Semenov (Eds.), *Water resources in Central Asia: International context* (pp. 103–120). Cham: Springer.

Smithers, J., & Blay-Palmer, A. (2001). Technology innovation as a strategy for climate adaptation in agriculture. *Applied Geography, 21,* 175–197.

Sorg, A., Bolch, T., Stoffel, M., Solomina, O., & Beniston, M. (2012). Climate change impacts on glaciers and runoff in Tien Shan (Central Asia). *Nature Climate Change, 2,* 725.

Spoor, M. (1998). The Aral Sea basin crisis: Transition and environment in former soviet Central Asia. *Development and Change, 29,* 409–435.

Wang, H. L., Gan, Y. T., Wang, R. Y., Niu, J. Y., Zhao, H., Yang, Q. G., et al. (2008). Phenological trends in winter wheat and spring cotton in response to climate changes in northwest China. *Agricultural and Forest Meteorology, 148,* 1242–1251.

Yue, S., Pilon, P., & Cavadias, G. (2002). Power of the Mann-Kendall and Spearman's rho tests for detecting monotonic trends in hydrological series. *Journal of Hydrology, 259,* 254–271.

Zonn, I. S., Kostianoy, A. G., Lokteva, T. V., & Shtol, V. V. (2018). Turkmenistan water resources policy in Central Asia. In *The handbook of environmental chemistry*. Berlin: Springer.

Zou, S., Jilili, A., Duan, W., Maeyer, P. D., & de Voorde, T. V. (2019). Human and natural impacts on the water resources in the Syr Darya River Basin, Central Asia. *Sustainability, 11,* 3084.

The manufacturer's authorised representative in the EU is Springer Nature Customer Service Centre GmbH, Europaplatz 3, 69115 Heidelberg, Germany. If you have any concerns regarding our products, please contact ProductSafety@springernature.com

Printed and bound by CPI Group (UK) Ltd, Croydon, CR0 4YY
30/09/2026
02225449-0002